I0045471

George Darwin

Problems Connected with the Tides of a Viscous Spheroid

George Darwin

Problems Connected with the Tides of a Viscous Spheroid

ISBN/EAN: 9783337213992

Printed in Europe, USA, Canada, Australia, Japan

Cover: Foto ©berggeist007 / pixelio.de

More available books at **www.hansebooks.com**

XIV. *Problems connected with the Tides of a Viscous Spheroid.*

By G. H. DARWIN, *M.A., Fellow of Trinity College, Cambridge.*

Communicated by J. W. L. GLAISHER, *M.A., F.R.S.*

Received November 14,—Read December 19, 1878.

CONTENTS.

IN the following paper several problems are considered, which were alluded to in my two previous papers on this subject.*

The paper is divided into sections which deal with the problems referred to in the table of contents. It was found advantageous to throw the several investigations together, because their separation would have entailed a good deal of repetition, and one system of notation now serves throughout.

It has, of course, been impossible to render the mathematical parts entirely independent of the previous papers, to which I shall accordingly have occasion to make a good many references.

As the whole inquiry is directed by considerations of applicability to the earth, I shall retain the convenient phraseology afforded by speaking of the tidally distorted spheroid as the earth, and of the disturbing body as the moon.

It is probable that but few readers will care to go through the somewhat complex arguments and analysis by which the conclusions are supported, and therefore in the fourth part a summary of results is given, together with some discussion of their physical applicability to the case of the earth.

I. *Secular distortion of the spheroid, and certain tides of the second order.*

In considering the tides of a viscous spheroid, it was supposed that the tidal protuberances might be considered as the excess and deficiency of matter above and below

* "On the Bodily Tides of Viscous and Semi-elastic Spheroids, and on the Ocean Tides upon a Yielding Nucleus," Phil. Trans., 1879, Part I., and—

"On the Precession of a Viscous Spheroid, and on the remote History of the Earth," immediately preceding the present paper. They will be referred to hereafter as "Tides" and "Precession" respectively.

the mean sphere—or more strictly the mean spheroid of revolution which represents the average shape of the earth. The spheroid was endued with the power of gravitation, and it was shown that the action of the spheroid on its own tides might be found approximately by considering the state of flow in the mean sphere caused by the attraction of the protuberances, and also by supposing the action of the protuberances on the sphere to be normal thereto, and to consist, in fact, merely of the weight (either positive or negative) of the protuberances.

Thus if a be the mean radius of the sphere, w its density, g mean gravity at the surface, and $r=a+\sigma_i$ the equation to the tidal protuberance, where σ_i is a surface harmonic of order i, the potential per unit volume of the protuberance in the interior of the sphere is $\frac{3gw}{2i+1}\left(\frac{r}{a}\right)^i\sigma_i$, and the sphere is subjected to a normal traction per unit area of surface equal to $-gw\sigma_i$.

It was also shown that these two actions might be compounded by considering the interior of the sphere (now free of gravitation) to be under the action of a potential $-\frac{2(i-1)}{2i+1}gw\left(\frac{r}{a}\right)^i\sigma_i$.

This expression therefore gave the effective potential when the sphere was treated as devoid of gravitational power.

It was remarked* that, strictly speaking, there is tangential action between the protuberance and the surface of the sphere. And later† it was stated that the action of an external tide-generating body on the lagging tides was not such as to form a rigorously equilibrating system of forces. The effects of this non-equilibration, in as far as it modifies the rotation of the spheroid as a whole, were considered in the paper on "Precession."

It is easy to see from general considerations that these previously neglected tangential stresses on the surface of the sphere, together with the effects of inertia due to the secular retardation of the earth's rotation (produced by the non-equilibrating forces), must cause a secular distortion of the spheroid.

This distortion I now propose to investigate.

In order to avoid unnecessary complication, the tides will be supposed to be raised by a single disturbing body or moon moving in the plane of the earth's equator.

Let $r=a+\sigma$ be the equation to the bounding surface of the tidally-distorted earth, where σ is a surface harmonic of the second order.

I shall now consider how the equilibrium is maintained of the layer of matter σ, as acted on by the attraction of the spheroid and under the influence of an external disturbing potential V, which is a solid harmonic of the second degree of the coordinates of points within the sphere.‡ The object to be attained is the evaluation of the stresses

* "Tides," Section 2.

† "Tides," Section 5.

‡ A parallel investigation would be applicable, where σ and V are of any orders.

tangential to the surface of the sphere, which are exercised by the layer σ on the sphere.

Let θ, ϕ be the colatitude and longitude of a point in the layer. Then consider a prismatic element bounded by the two cones θ, $\theta+\delta\theta$, and by the two planes ϕ, $\phi+\delta\phi$.

The radial faces of this prism are acted on by the pressures and tangential stresses communicated by the four contiguous prisms. But the tangential stresses on these faces only arise from the fact that contiguous prisms are solicited by slightly different forces, and therefore the action of the four prisms, surrounding the prism in question, must be principally pressure. I therefore propose to consider that the prism resists the tendency of the impressed forces to move tangentially along the surface of the sphere, by means of hydrostatic pressures on its four radial faces, and by a tangential stress across its base.

This approximation by which the whole of the tangential stress is thrown on to the base, is clearly such as slightly to accentuate, as it were, the distribution of the tangential stresses on the surface of the sphere, by which the equilibrium of the layer σ is maintained. For consider the following special case :—Suppose σ to be a surface of revolution, and V to be such that only a single small circle of latitude is solicited by a tangential force everywhere perpendicular to the meridian. Then it is obvious that, strictly speaking, the elements lying a short way north and south of the small circle would tend to be carried with it, and the tangential stress on the sphere would be a maximum along the small circle, and would gradually die away to the north and south. In the approximate method, however, which it is proposed to use, such an application of external force would be deemed to cause no tangential stress to the surface of the sphere to the north and south of the small circle acted on. This special case is clearly a great exaggeration of what holds in our problem, because it postulates a finite difference of disturbing force between elements infinitely near to one another.

We will first find what are the hydrostatic pressures transmitted by the four prisms contiguous to the one we are considering.

Let p be the hydrostatic pressure at the point r, θ, ϕ of the layer σ. Then if we neglect the variations of gravity due to the layer σ and to V, p is entirely due to the attraction of the mean sphere of radius a.

The mean pressure on the radial faces at the point in question is $\frac{1}{2}gw\sigma$; where σ is negative the pressures are of course tractions.

We will first resolve along the meridian.

The excess of the pressure acting on the face $\theta+\delta\theta$ over that on the face θ (whose area is $\sigma a \sin\theta\delta\phi$) is

$$\frac{d}{d\theta}[\tfrac{1}{2}gw\sigma . \sigma a \sin\theta\delta\phi]\delta\theta, \text{ or } \tfrac{1}{2}gwa\frac{d}{d\theta}(\sigma^2 \sin\theta)\delta\theta\delta\phi,$$

and it acts towards the pole.

The resolved part of the pressures on the faces $\phi+\delta\phi$ and ϕ (whose area is $\sigma a\delta\theta$) along the meridian is

$$(\tfrac{1}{2}gw\sigma)(\sigma a\delta\theta)(\cos\theta\delta\phi) \text{ or } \tfrac{1}{2}gwa\sigma^2\cos\theta\delta\theta\delta\phi,$$

and it acts towards the equator.

Hence the whole force due to pressure on the element resolved along the meridian towards the equator is

$$\tfrac{1}{2}gwa\delta\theta\delta\phi\left(\sigma^2\cos\theta-\frac{d}{d\theta}(\sigma^2\sin\theta)\right), \text{ or } -gwa\delta\theta\delta\phi\sin\theta\sigma\frac{d\sigma}{d\theta}.$$

But the mass of the elementary prism $\delta m = wa^2\sin\theta\delta\theta\delta\phi.\sigma$.

Hence the meridional force due to pressure is $-\frac{g}{a}\delta m\frac{d\sigma}{d\theta}$.

We will next resolve the pressures perpendicular to the meridian.

The excess of pressure on the face $\phi+\delta\phi$ over that on the face ϕ (whose area is $\sigma a\delta\theta$), measured in the direction of ϕ increasing, is

$$-\frac{d}{d\phi}[\tfrac{1}{2}gw\sigma.\sigma a\delta\theta]\delta\phi = -gwa\sigma\frac{d\sigma}{d\phi}\delta\theta\delta\phi = -\frac{g}{a}\delta m\frac{1}{\sin\theta}\frac{d\sigma}{d\phi}.$$

Hence the force due to pressure perpendicular to the meridian is $-\frac{g}{a}\delta m\frac{1}{\sin\theta}\frac{d\sigma}{d\phi}$.

We have now to consider the impressed forces on the element.

Since σ is a surface harmonic of the second degree, the potential of the layer of matter σ at an external point is $\frac{3}{5}g\sigma\left(\frac{a}{r}\right)^3$. Therefore the forces along and perpendicular to the meridian on a particle of mass δm, just outside the layer σ but infinitely near the prismatic element, are $\frac{3}{5}\frac{g}{a}\delta m\frac{d\sigma}{d\theta}$ and $\frac{3}{5}\frac{g}{a}\delta m\frac{1}{\sin\theta}\frac{d\sigma}{d\phi}$, and these are also the forces acting on the element δm due to the attraction of the rest of the layer σ.

Lastly, the forces due to the external potential V are clearly $\delta m\frac{1}{a}\frac{dV}{d\theta}$ and $\delta m\frac{1}{a\sin\theta}\frac{dV}{d\phi}$.

Then collecting results we get for the forces due both to pressure and attraction, along the meridian towards the equator

$$\delta m\left[-\frac{g}{a}\frac{d\sigma}{d\theta}+\frac{3}{5}\frac{g}{a}\frac{d\sigma}{d\theta}+\frac{dV}{ad\theta}\right]=\delta m\frac{d}{ad\theta}(V-\tfrac{2}{5}g\sigma),$$

and perpendicular to the meridian, in the direction of ϕ increasing,

$$\delta m\left[-\frac{g}{a\sin\theta}\frac{d\sigma}{d\phi}+\frac{3g}{5a\sin\theta}\frac{d\sigma}{d\phi}+\frac{dV}{a\sin\theta d\phi}\right]=\delta m\frac{1}{a\sin\theta}\frac{d}{d\phi}(V-\tfrac{2}{5}g\sigma).$$

Henceforward $\frac{2g}{5a}$ will be written $\mathfrak{g}$, as in the previous papers.

Now these are the forces on the element which must be balanced by the tangential stresses across the base of the prismatic element.

It follows from the above formulas that the tangential stresses communicated by the layer σ to the surface of the sphere are those due to a potential $V-\mathfrak{g}a\sigma$ acting on the layer σ.

If $\sigma=\frac{V}{\mathfrak{g}a}$ there is no tangential stress. But this is the condition that σ should be the equilibrium tidal spheroid due to V, so that the result fulfils the condition that if σ be the equilibrium tidal spheroid of V there is no tendency to distort the spheroid further; this obviously ought to be the case.

In the problem before us, however, σ does not fulfil this condition, and therefore there is tangential stress across the base of each prismatic element tending to distort the sphere.

Suppose $V=r^2S$ where S is a surface harmonic.

Then at the surface $V=a^2S$. If δm be the mass of a prism cut out of the layer σ, which stands on unit area as base, then $\delta m=w\sigma$.

Therefore the tangential stresses per unit area communicated to the sphere are

$$\left.\begin{array}{ll} & wa^2\frac{\sigma}{a}\frac{d}{d\theta}\left(S-\mathfrak{g}\frac{\sigma}{a}\right) \text{ along the meridian} \\ \text{and} & \\ & wa^2\frac{\sigma}{a}\frac{1}{\sin\theta}\frac{d}{d\phi}\left(S-\mathfrak{g}\frac{\sigma}{a}\right) \text{ perpendicular to the meridian} \end{array}\right\} \quad . \quad (1)$$

Besides these tangential stresses there is a small radial stress over and above the radial traction $-gw\sigma$, which was taken into account in forming the tidal theory. But we remark that the part of this stress, which is periodic in time, will cause a very small tide of the second order, and the part which is non-periodic will cause a very small permanent modification of the figure of the sphere. But these effects are so minute as not to be worth investigating.

We will now apply these results to the tidal problem.

Let XYZ (fig. 1) be rectangular axes fixed in the earth, Z being the axis of rotation and XZ the plane from which longitudes are measured.

Let M be the projection of the moon on the equator, and let ω be the earth's angular velocity of rotation relatively to the moon.

Let A be the major axis of the tidal ellipsoid.

Let $AX=\omega t$, where t is the time, and let $MA=\epsilon$.

Let m be the moon's mass measured astronomically, and c her distance, and $\tau=\frac{3}{2}\frac{m}{c^3}$.

Then according to the usual formula, the moon's tide-generating potential is

$$\tau r^2[\sin^2\theta\cos^2(\phi-\omega t-\epsilon)-\tfrac{1}{3}],$$

which may be written

$$\tfrac{1}{2}\tau r^2(\tfrac{1}{3}-\cos^2\theta)+\tfrac{1}{2}\tau r^2\sin^2\theta\cos 2(\phi-\omega t-\epsilon).$$

The former of these terms is not a function of the time, and its effect is to cause a permanent small increase of ellipticity of figure of the earth, which may be neglected. We are thus left with

$$\tfrac{1}{2}\tau r^2\sin^2\theta\cos 2(\phi-\omega t-\epsilon)$$

as the true tide-generating potential.

Now if $\tan 2\epsilon=\frac{19\upsilon\omega}{\mathfrak{g}a w}$, where υ is the coefficient of viscosity of the spheroid, then by the theory of the paper on "Tides," such a potential will raise a tide expressed by

$$\frac{\sigma}{a}=\frac{1}{2}\frac{\tau}{\mathfrak{g}}\cos 2\epsilon\sin^2\theta\cos 2(\phi-\omega t)^* \quad . \tag{2}$$

Then if we put

$$S=\tfrac{1}{2}\tau\sin^2\theta\cos 2(\phi-\omega t-\epsilon)\,. \quad . \quad . \quad . \quad . \quad . \quad . \tag{3}$$

$$S-\mathfrak{g}\frac{\sigma}{a}=\tfrac{1}{2}\tau\sin 2\epsilon\sin^2\theta\sin 2(\phi-\omega t) \quad . \quad . \quad . \quad . \quad . \quad . \quad . \tag{4}$$

and

$$\frac{d}{d\theta}\left(S-\mathfrak{g}\frac{\sigma}{a}\right)=\tau\sin 2\epsilon\sin\theta\cos\theta\sin 2(\phi-\omega t)$$

$$\frac{1}{\sin\theta}\frac{d}{d\phi}\left(S-\mathfrak{g}\frac{\sigma}{a}\right)=\tau\sin 2\epsilon\sin\theta\cos 2(\phi-\omega t).$$

Multiplying these by $wa^2\frac{\sigma}{a}$, we find from (1) the tangential stresses communicated by the layer σ to the sphere.

* "Tides," Section 5.

They are

$$wa^2 \tfrac{1}{8}\frac{\tau^2}{\mathfrak{g}} \sin 4\epsilon \sin^3\theta \cos\theta \sin 4(\phi-\omega t) \text{ along the meridian,}$$

and

$$wa^2 \tfrac{1}{8}\frac{\tau^2}{\mathfrak{g}} \sin 4\epsilon \sin^3\theta (1+\cos 4(\phi-\omega t)) \text{ perpendicular to the meridian.}$$

These stresses of course vanish when ϵ is zero, that is to say when the spheroid is perfectly fluid.

In as far as they involve $\phi-\omega t$ these expressions are periodic, and the periodic parts must correspond with periodic inequalities in the state of flow of the interior of the earth. These small tides of the second order have no present interest and may be neglected.

We are left, therefore, with a non-periodic tangential stress per unit area of the surface of the sphere perpendicular to the meridian from east to west equal to $\tfrac{1}{8}wa^2\frac{\tau^2}{\mathfrak{g}} \sin 4\epsilon \sin^3\theta$.

The sum of the moments of these stresses about the axis Z constitutes the tidal frictional couple $\mathfrak{N}$, which retards the earth's rotation.

Therefore

$$\mathfrak{N} = \tfrac{1}{8}wa^2\frac{\tau^2}{\mathfrak{g}} \sin 4\epsilon \iint \sin^3\theta . a \sin\theta . a^2 \sin\theta d\theta d\phi$$

integrated all over the surface of the sphere, and effecting the integration we have $\mathfrak{N} = \frac{4\pi}{15} wa^5 . \frac{\tau^2}{\mathfrak{g}} \sin 4\epsilon$.

But if C be the earth's moment of inertia, $C = \frac{8}{15}\pi wa^5$.

Therefore

$$\frac{\mathfrak{N}}{C} = \tfrac{1}{2}\frac{\tau^2}{\mathfrak{g}} \sin 4\epsilon \quad . \qquad (5)$$

This expression agrees with that found by a different method in the paper on "Precession."*

We may now write the tangential stress on the surface of the sphere as $\tfrac{1}{4}wa^2\frac{\mathfrak{N}}{C}\sin^3\theta$; and the components of this stress parallel to the axes X, Y, Z are

$$-\tfrac{1}{4}wa^2\frac{\mathfrak{N}}{C}\sin^3\theta \sin\phi, \; +\tfrac{1}{4}wa^2\frac{\mathfrak{N}}{C}\sin^3\theta\cos\phi, \; 0 \quad . \; . \; . \; . \; . \qquad (6)$$

We now have to consider those effects of inertia which equilibrate this system of surface forces.

The couple $\mathfrak{N}$ retards the earth's rotation very nearly as though it were a rigid

* "Precession," Section 5 (22), when $i=0$.

body. Hence the effective force due to inertia on a unit of volume of the interior of the earth at a point r, θ, ϕ is $wr \sin \theta \frac{\mathfrak{N}}{C}$, and it acts in a small circle of latitude from west to east. The sum of the moments of these forces about the axis of Z is of course equal to $\mathfrak{N}$, and therefore this bodily force would equilibrate the surface forces found in (6), if the earth were rigid.

The components of the bodily force parallel to the axes are in rectangular co-ordinates.

$$wy\frac{\mathfrak{N}}{C},\ -wx\frac{\mathfrak{N}}{C},\ 0 \quad \ldots \quad (7)$$

The problem is therefore reduced to that of finding the state of flow in the interior of a viscous sphere, which is subject to a bodily force of which the components are (7) and to the surface stresses of which the components are (6).

Let α, β, γ be the component velocities of flow at the point x, y, z, and υ the coefficient of viscosity. Then neglecting inertia because the motion is very slow, the equations of motion are

$$\left.\begin{aligned} -\frac{dp}{dx}+\upsilon\nabla^2\alpha+w\frac{\mathfrak{N}}{C}y&=0\\ -\frac{dp}{dy}+\upsilon\nabla^2\beta-w\frac{\mathfrak{N}}{C}x&=0\\ -\frac{dp}{dz}+\upsilon\nabla^2\gamma&=0\\ \frac{d\alpha}{dx}+\frac{d\beta}{dy}+\frac{d\gamma}{dz}&=0 \end{aligned}\right\} \quad . \quad (8)$$

We have to find a solution of these equations, subject to the condition above stated, as to surface stress.

Let α', β', γ', p' be functions which satisfy the equations (8) throughout the sphere. Then if we put $\alpha=\alpha'+\alpha_{,}$, $\beta=\beta'+\beta_{,}$, $\gamma=\gamma'+\gamma_{,}$, $p=p'+p_{,}$ we see that to complete the solution we have to find $\alpha_{,}$, $\beta_{,}$, $\gamma_{,}$, $p_{,}$ as determined by the equations

$$\frac{dp_{,}}{dx}+\upsilon\nabla^2\alpha_{,}=0,\ \frac{dp_{,}}{dy}\ \&c.=0,\ \frac{dp_{,}}{dz}\ \&c.=0,\ \frac{d\alpha_{,}}{dx}+\frac{d\beta_{,}}{dy}+\frac{d\gamma_{,}}{dz}=0 \quad . \quad . \quad . \quad . \quad (9)$$

which they are to satisfy throughout the sphere. They must also satisfy certain equations to be found by subtracting from the given surface stresses (6), components of surface stress to be calculated from α', β', γ', p'.*

We have first to find α', β', γ', p'.

Conceive the symbols in equations (8) to be accented, and differentiate the first

* This statement of method is taken from THOMSON and TAIT'S 'Nat. Phil.,' § 733.

three by x, y, z respectively and add them; then bearing in mind the fourth equation, we have $\nabla^2 p'=0$, of which $p'=0$ is a solution.

Thus the equations to be satisfied become

$$\nabla^2\alpha'=-\frac{w}{v}\frac{\mathfrak{N}}{C}y,\ \nabla^2\beta'=\frac{w}{v}\frac{\mathfrak{N}}{C}x,\ \nabla^2\gamma'=0.$$

Solutions of these are obviously

$$\left.\begin{aligned}\alpha'&=-\tfrac{1}{10}\frac{w}{v}\frac{\mathfrak{N}}{C}r^2y, & \beta'&=\tfrac{1}{10}\frac{w}{v}\frac{\mathfrak{N}}{C}r^2x, & \gamma'&=0\\ &=-\tfrac{1}{10}\frac{w}{v}\frac{\mathfrak{N}}{C}r^3\sin\theta\sin\phi & &=\tfrac{1}{10}\frac{w}{v}\frac{\mathfrak{N}}{C}r^3\sin\theta\cos\phi & &\end{aligned}\right\} \quad . \ . \ . \ . \quad (10)$$

These values satisfy the last of (8), viz.: the equation of continuity, and therefore together with $p'=0$, they form the required values of α', β', γ', p'.

We have next to compute the surface stresses corresponding to these values.

Let P, Q, R, S, T, U be the normal and tangential stresses (estimated as is usual in the theory of elastic solids) across three planes at right angles at the point x, y, z.

Then

$$P=-p'+2v\frac{d\alpha'}{dx},\ S=v\left(\frac{d\beta'}{dz}+\frac{d\gamma'}{dy}\right) \quad . \ . \ . \ . \ . \ . \ . \ . \quad (11)$$

Q, R, T, U being found by cyclical changes of symbols.

Let F, G, H be the component stresses across a plane perpendicular to the radius vector r at the point x, y, z; then

$$\left.\begin{aligned}Fr&=Px+Uy+Tz\\ Gr&=Ux+Qy+Sz\\ Hr&=Tx+Sy+Rz\end{aligned}\right\} \quad . \ . \ . \ . \ . \quad (12)$$

Substitute in (12) for P, Q, &c., from (11), and put $\zeta'=\alpha'x+\beta'y+\gamma'z$, and $r\frac{d}{dr}$ for $x\frac{d}{dx}+y\frac{d}{dy}+z\frac{d}{dz}$. Then

$$Fr=-p'x+v\left\{\left(r\frac{d}{dr}-1\right)\alpha'+\frac{d\zeta'}{dx}\right\},\ Gr=\&c.,\ Hr=\&c. \quad . \ . \ . \ . \ . \quad (13)$$

These formulas give the stresses across any of the concentric spherical surfaces.

In the particular case in hand $p'=0$, $\gamma'=0$, $\zeta'=0$, and α', β' are homogeneous functions of the third degree, hence

$$F=-\tfrac{1}{5}w\frac{\mathfrak{N}}{C}r^2\sin\theta\sin\phi,\quad G=\tfrac{1}{5}w\frac{\mathfrak{N}}{C}r^2\sin\theta\cos\phi,\quad H=0\ . \qquad (14)$$

and at the surface of the sphere $r=a$.

Then according to the principles above explained, we have to find $\alpha_{,}$, $\beta_{,}$, $\gamma_{,}$ so that they may satisfy

$$-\frac{dp_{,}}{dx}+\upsilon\nabla^2\alpha_{,}=0,\ \&c.,\ \&c.,$$

throughout a sphere, which is subject to surface stresses given by subtracting from (6) the surface values of F, G, H in (14). Hence the surface stresses to be satisfied by $\alpha_{,}$, $\beta_{,}$, $\gamma_{,}$ have components

$$A_3=\frac{w}{4}\frac{\mathfrak{N}}{C}a^2(\tfrac{4}{5}-\sin^2\theta)\sin\theta\sin\phi,\quad B_3=-\frac{w}{4}\frac{\mathfrak{N}}{C}a^2(\tfrac{4}{5}-\sin^2\theta)\sin\theta\cos\phi,\quad C_3=0.$$

These are surface harmonics of the third order as they stand.

Now the solution of Sir W. THOMSON's problem of the state of strain of an incompressible elastic sphere, subject only to surface stress, is applicable to an incompressible viscous sphere, *mutatis mutandis.* His solution* shows that a surface stress, of which the components are A_i, B_i, C_i (surface harmonics of the ith order), gives rise to a state of flow expressed by

$$\alpha=\frac{1}{\upsilon a^{i-1}}\left\{\frac{(a^2-r^2)}{2(2i^2+1)}\frac{d\Psi_{i-1}}{dx}+\frac{1}{i-1}\left[\frac{(i+2)r^{2i+1}}{(2i^2+1)(2i+1)}\frac{d}{dx}(\Psi_{i-1}r^{-2i+1})+\frac{1}{2i(2i+1)}\frac{d\Phi_{i+1}}{dx}+A_ir^i\right]\right\} \qquad (15)$$

and symmetrical expressions for β, γ.

Where Ψ and Φ are auxiliary functions defined by

$$\left.\begin{aligned}\Psi_{i-1}&=\frac{d}{dx}(A_ir^i)+\frac{d}{dy}(B_ir^i)+\frac{d}{dz}(C_ir^i)\\ \Phi_{i+1}&=r^{2i+3}\left\{\frac{d}{dx}(A_ir^{-i-1})+\frac{d}{dy}(B_ir^{-i-1})+\frac{d}{dz}(C_ir^{-i-1})\right\}\end{aligned}\right\} \qquad (16)$$

In our case $i=3$, and it is easily shown that the auxiliary functions are both zero, so that the required solution is

$$\alpha_{,}=\frac{w}{8\upsilon}\frac{\mathfrak{N}}{C}(\tfrac{4}{5}-\sin^2\theta)r^3\sin\theta\sin\phi,\quad \beta_{,}=-\frac{w}{8\upsilon}\frac{\mathfrak{N}}{C}(\tfrac{4}{5}-\sin^2\theta)\sin\theta\cos\phi,\quad \gamma_{,}=0.$$

If we add to these the values of α', β', γ' from (10), we have as the complete solution of the problem,

* THOMSON and TAIT's 'Nat. Phil.,' § 737.

$$\alpha=-\frac{w}{8v}\frac{\mathfrak{P}}{C}r^3\sin^3\theta\sin\phi,\ \beta=\frac{w}{8v}\frac{\mathfrak{P}}{C}r^3\sin^3\theta\cos\phi,\ \gamma=0\quad .\quad .\quad .\quad .\quad (17)$$

These values show that the motion is simply cylindrical round the earth's axis, each point moving in a small circle of latitude from east to west with a linear velocity $\frac{w}{8v}\frac{\mathfrak{P}}{C}r^3\sin^3\theta$, or with an angular velocity about the axis equal to $\frac{w}{8v}\frac{\mathfrak{P}}{C}r^2\sin^2\theta$.*

In this statement a meridian at the pole is the curve of reference, but it is more intelligible to state that each particle moves from west to east with an angular velocity about the axis equal to $\frac{w}{8v}\frac{\mathfrak{P}}{C}(a^2-r^2\sin^2\theta)$, with reference to a point on the surface at the equator.

The easterly rate of change of the longitude L of any point on the surface in colatitude θ is therefore $\frac{wa^2}{8v}\frac{\mathfrak{P}}{C}\cos^2\theta$.

Then since $\frac{\mathfrak{P}}{C}=\frac{\tau^2}{\mathfrak{g}}\sin 2\epsilon\cos 2\epsilon$, and $\tan 2\epsilon=\frac{2}{5}\cdot\frac{19v\omega}{\mathfrak{g}wa^2}$, therefore

$$\frac{dL}{dt}=\tfrac{19}{20}\left(\frac{\tau}{\mathfrak{g}}\cos 2\epsilon\right)^2\omega\cos^2\theta \quad (17')$$

This equation gives the rate of change of longitude. The solution is not applicable to the case of perfect fluidity, because the terms introduced by inertia in the equations of motion have been neglected; and if the viscosity be infinitely small, the inertia terms are no longer small compared with those introduced by viscosity.

In order to find the total change of longitude in a given period, it will be more convenient to proceed from a different formula.

Let n, Ω be the earth's rotation, and the moon's orbital motion at any time; and let the suffix 0 to any symbol denote its initial value, also let $\xi=\left(\frac{\Omega_0}{\Omega}\right)^{\frac{1}{3}}$.

Then it was shown in the paper on "Precession" that the equation of conservation of moment of momentum of the moon-earth system is

$$\frac{n}{n_0}=1+\mu(1-\xi)\dagger\quad .\quad .\quad .\quad (18)$$

Where μ is a certain constant, which in the case of the homogeneous earth with the present lengths of day and month, is almost exactly equal to 4.

By differentiation of (18)

$$\frac{dn}{dt}=-\mu n_0\frac{d\xi}{dt}\quad .\quad (19)$$

* The problem might probably be solved more shortly without using the general solution, but the general solution will be required in Part III.

† "Precession," equation (73), when $i=0$ and $\tau'=0$.

But the equation of tidal friction is $\frac{dn}{dt}=-\frac{\mathfrak{N}}{C}$. Therefore

$$\frac{d\xi}{dt}=\frac{1}{\mu}\frac{\mathfrak{N}}{Cn_0}$$

Now

$$\frac{dL}{dt}=\frac{wa^2}{8v}\frac{\mathfrak{N}}{C}\cos^2\theta.$$

Therefore

$$\frac{dL}{d\xi}=\mu n_0\frac{wa^2}{8v}\cos^2\theta. \quad . \quad . \quad . \quad . \quad . \quad . \quad . \qquad (19')$$

All the quantities on the right-hand side of this equation are constant, and therefore by integration we have for the change of longitude

$$\Delta L=\mu n_0\frac{wa^2}{8v}(\xi-1)\cos^2\theta.$$

But since $\omega_0=n_0-\Omega_0$, and $\tan 2\epsilon_0=\frac{2}{5}\cdot\frac{19v\omega_0}{\mathfrak{g}wa^2}$, therefore in degrees of arc,

$$\Delta L=\frac{180}{\pi}\mu n_0\frac{19}{20}\frac{n_0-\Omega_0}{\mathfrak{g}}\cot 2\epsilon_0(\xi-1)\cos^2\theta.$$

In order to make the numerical results comparable with those in the paper on "Precession," I will apply this to the particular case which was the subject of the first method of integration of that paper.* It was there supposed that $\epsilon_0=17^\circ\ 30'$, and it was shown that looking back about 46 million years ξ had fallen from unity to ·88. Substituting for the various quantities their numerical values, I find that

$$-\Delta L=0^\circ{\cdot}31\cos^2\theta=19'\cos^2\theta.$$

Hence looking back 46 million years, we find the longitude of a point in latitude 30°, further west by $4\frac{3}{4}'$ than at present, and a point in latitude 60°, further west by $14\frac{1}{4}'$—both being referred to a point on the equator.

Such a shift is obviously quite insignificant, but in order to see whether this screwing motion of the earth's mass could have had any influence on the crushing of the surface strata, it will be well to estimate the amount by which a cubic foot of the earth's mass at the surface would have been distorted.

The motion being referred to the pole, it appears from (17) that a point distant ρ from the axis shifts through $\frac{w}{8v}\frac{\mathfrak{N}}{C}\rho^3\delta t$ in the time δt. There would be no shearing if

* "Precession," Section 15.

a point distant $\rho+\delta\rho$ shifted through $\frac{w}{8v}\frac{\mathfrak{N}}{C}\rho^2(\rho+\delta\rho)\delta t$; but this second point does shift through $\frac{w}{8v}\frac{\mathfrak{N}}{C}(\rho+\delta\rho)^3\delta t$.

Hence the amount of shear in unit time is

$$\frac{1}{\delta\rho}\times\frac{w}{8v}\frac{\mathfrak{N}}{C}\left[(\rho+\delta\rho)^3-(\rho+\delta\rho)\rho^2\right]=\frac{w}{4v}\frac{\mathfrak{N}}{C}\rho^2.$$

Therefore at the equator, at the surface where the shear is greatest, the shear per unit time is

$$\frac{wa^2}{4v}\frac{\mathfrak{N}}{C}=\tfrac{19}{10}\left(\frac{\tau}{\mathfrak{g}}\right)^2\cos^2 2\epsilon\,.\,\omega.$$

With the present values of τ and ω, $\frac{19}{10}\left(\frac{\tau}{\mathfrak{g}}\right)^2\omega$ is a shear of $\frac{1\cdot 84}{10^{10}}$ per annum.

Hence at the equator a slab one foot thick would have one face displaced with reference to the other at the rate of $\frac{1}{500}\cos^2 2\epsilon$ of an inch in a million years.

The bearing of these results on the history of the earth will be considered in Part IV.

The next point which will be considered is certain tides of the second order.

We have hitherto supposed that the tides are superposed upon a sphere; it is, however, clear that besides the tidal protuberance there is a permanent equatorial protuberance. Now this permanent protuberance is by hypothesis not rigidly connected with the mean sphere; and, as the attraction of the moon on the equatorial regions produces the uniform precession and the fortnightly nutation, it might be (and indeed has been) supposed that there would arise a shifting of the surface with reference to the interior, and that this change in configuration would cause the earth to rotate round a new axis, and so there would follow a geographical shifting of the poles. I will now show, however, that the only consequence of the non-rigid attachment of the equatorial protuberance to the mean sphere is a series of tides of the second order in magnitude, and of higher orders of harmonics than the second.

For a complete solution of the problem the task before us would be to determine what are the additional tangential and normal stresses existing between the protuberant parts and the mean sphere, and then to find the tides and secular distortion (if any) to which they give rise.

The first part of these operations may be done by the same process which has just been carried out with reference to the secular distortion due to tidal friction.

The additional normal stress (in excess of $-gw\sigma$, the mean weight of an element of the protuberance) can have no part in the precessional and nutational couples, and the remark may be repeated that, that part of it which is non-periodic will only cause a minute change in the mean figure of the spheroid which is negligeable, and the part which is periodic will cause small tides of about the same magnitude as those caused by the tangential stresses. With respect to the tangential stresses, it is *à priori* possible that they may cause a continued distortion of the spheroid, and they will cause certain small tides, whose relative importance we have to estimate.

The expressions for the tangential stresses, which we have found above in (1), are not linear, and therefore we must consider the phenomenon in its entirety, and must not seek to consider the precessional and nutational effects apart from the tidal effects.

The whole bodily potential which acts on the earth is that due to the moon (of which the full expression is given in equation (3) of "Precession"), together with that due to the earth's diurnal rotation (being $\frac{1}{2}n^2r^2(\frac{1}{3}-\cos^2\theta)$); the whole may be called r^2S. The form of the surface σ is that due to the tides and to the non-periodic part of the moon's potential, together with that due to rotation—being $\frac{a}{2}\frac{n^2}{\mathfrak{g}}(\frac{1}{3}-\cos^2\theta)$.

Now if we form the effective potential $a^2\left(\mathrm{S}-\mathfrak{g}\frac{\sigma}{a}\right)$, which determines the tangential stresses between σ and the mean sphere, we shall find that all except periodic terms disappear. This is so whether we suppose the earth's axis to be oblique or not to the lunar orbit, and also if the sun be supposed to act.

Then if we differentiate these and form the expressions

$$wa^2\frac{\sigma}{a}\frac{d}{d\theta}\left(\mathrm{S}-\mathfrak{g}\frac{\sigma}{a}\right),\quad wa^2\frac{\sigma}{a}\frac{d}{\sin\theta d\phi}\left(\mathrm{S}-\mathfrak{g}\frac{\sigma}{a}\right),$$

we shall find that there are no non-periodic terms in the expression giving the tangential stress along the meridian; and that the only non-periodic terms which exist in the expression giving the tangential stress perpendicular to the meridian are precisely those whose effects have been already considered as causing secular distortion, and which have their maximum effect when the obliquity is zero.

Hence the whole result must be—

(1) A very minute change in the permanent or average figure of the globe;

(2) The secular distortion already investigated;

(3) Small tides of the second order.

The one question which is of interest is, therefore—Can these small tides be of any importance?

The sum of the moments of all the tangential stresses which result from the above expressions, about a pair of axes in the equator, one 90° removed from the moon's meridian and the other in the moon's meridian, together give rise to the precessional and nutational couples.

Hence it follows that part of the tangential stresses form a non-equibrating system of forces acting on the sphere's surface. In order to find the distorting effects on the globe, we should, therefore, have to equibrate the system by bodily forces arising from the effects of the inertia due to the uniform precession and the fortnightly nutation—just as was done above with the tidal friction. This would be an exceedingly laborious process; and although it seems certain that the tides thus raised would be very small, yet we are fortunately able to satisfy ourselves of the fact more rigorously. Certain parts of the tangential stresses *do* form an equibrating system of forces, and these are precisely those parts of the stresses which are the most important, because they do not involve the sine of the obliquity.

I shall therefore evaluate the tangential stresses when the obliquity is zero.

The complete potential due both to the moon and to the diurnal rotation is

$$r^2\mathrm{S}=\tfrac{1}{2}r^2(n^2+\tau)(\tfrac{1}{3}-\cos^2\theta)+\tfrac{1}{2}r^2\tau\sin^2\theta\cos 2(\phi-\omega t-\epsilon),$$

and the complete expression for the surface of the spheroid is given by

$$\mathfrak{g}\frac{\sigma}{a}=\tfrac{1}{2}(n^2+\tau)(\tfrac{1}{3}-\cos^2\theta)+\tfrac{1}{2}\tau\cos 2\epsilon\sin^2\theta\cos 2(\phi-\omega t).$$

Hence

$$\mathrm{S}-\mathfrak{g}\frac{\sigma}{a}=\tfrac{1}{2}\tau\sin 2\epsilon\sin^2\theta\sin 2(\phi-\omega t).$$

Then neglecting τ^2 compared with τn^2, and omitting the terms which were previously considered as giving rise to secular distortion, we find

$$wa^2\frac{\sigma}{a}\frac{d}{d\theta}\left(\mathrm{S}-\mathfrak{g}\frac{\sigma}{a}\right)=wa^2\tau\tfrac{1}{2}\frac{n^2}{\mathfrak{g}}\sin 2\epsilon\sin\theta\cos\theta(\tfrac{1}{3}-\cos^2\theta)\sin 2(\phi-\omega t),$$

$$wa^2\frac{\sigma}{a}\frac{d}{\sin\theta d\phi}\left(\mathrm{S}-\mathfrak{g}\frac{\sigma}{a}\right)=wa^2\tau\tfrac{1}{2}\frac{n^2}{\mathfrak{g}}\sin 2\epsilon\sin\theta(\tfrac{1}{3}-\cos^2\theta)\cos 2(\phi-\omega t).$$

The former gives the tangential stress along, and the latter perpendicular to, the meridian.

If we put $e=\tfrac{1}{2}\frac{n^2}{\mathfrak{g}}$, the ellipticity of the spheroid, we see that the intensity of the tangential stresses is estimated by the quantity $wa^2.\tau e\sin 2\epsilon$. But we must now find a standard of comparison, in order to see what height of tide such stresses would be competent to produce.

It appears from a comparison of equations (7) and (8) of Section 2 of the paper on "Tides," that a surface traction S_i (a surface harmonic) everywhere normal to the sphere produces the same state of flow as that caused by a bodily force, whose potential per unit volume is $\left(\frac{r}{a}\right)^i\mathrm{S}_i$; and conversely a potential W_i is mechanically equivalent to a surface traction $\left(\frac{a}{r}\right)^i\mathrm{W}_i$.

Now the tides of the first order are those due to an effective potential $wr^2\left(\mathrm{S}-\mathfrak{g}\frac{\sigma}{a}\right)$, and hence the surface normal traction which is competent to produce the tides of the first order is $wa^2\left(S-\mathfrak{g}\frac{\sigma}{a}\right)$, which is equal to $wa^2\frac{1}{2}\tau\sin 2\epsilon\sin^2\theta\sin 2(\phi-\omega t)$. Hence the intensity of this normal traction is estimated by the quantity $wa^2\frac{1}{2}\tau\sin 2\epsilon$, and this affords a standard of comparison with the quantity $wa^2\tau e\sin 2\epsilon$, which was the estimate of the intensity of the secondary tides. The ratio of the two is $2e$, and since the ellipticity of the mean spheroid is small, the secondary tides must be small compared with the primary ones. It cannot be asserted that the ratio of the heights of the two tides will be $2e$, because the secondary tides are of a higher order of harmonics than the primary, and because the tangential stresses have not been reduced to harmonics and the problem completely worked out. I think it probable that the height of the secondary tides would be considerably less than is expressed by the quantity $2e$, but all that we are concerned to know is that they will be negligeable, and this is established by the preceding calculations.

It follows, then, that the precessional and nutational forces will cause no secular shifting of the surface with reference to the interior, and therefore cannot cause any such geographical deplacement of the poles, as has been sometimes supposed.

II. *The distribution of heat generated by internal friction and secular cooling.*

In the paper on "Precession" (Section 16) the total amount of heat was found, which was generated in the interior of the earth, in the course of its retardation by tidal friction. The investigation was founded on the principle that the energy, both kinetic and potential, of the moon-earth system, which was lost during any period, must reappear as heat in the interior of the earth. This method could of course give no indication of the manner and distribution of the generation of heat in the interior. Now the distribution of heat must have a very important influence on the way it will affect the secular cooling of the earth's mass, and I therefore now propose to investigate the subject from a different point of view.

It will be sufficient for the present purpose if we suppose the obliquity to the ecliptic to be zero, and the earth to be tidally distorted by the moon alone.

It has already been explained in the first section how we may neglect the mutual gravitation of a spheroid tidally distorted by an external disturbing potential wr^2S, if we suppose the disturbing potential to be $wr^2\left(S-\mathfrak{g}\frac{\sigma}{a}\right)$, where $r=a+\sigma$ is the equation to the tidal protuberance.

It is shown in (4) that

$$S-\mathfrak{g}\frac{\sigma}{a}=\frac{1}{2}\tau\sin 2\epsilon\sin^2\theta\sin 2(\phi-\omega t).$$

If we refer the motion to rectangular axes rotating so that the axis of x is the major axis of the tidal spheroid, and that of z is the earth's axis of rotation, and if W be the effective disturbing potential estimated per unit volume, we have

$$\mathrm{W}=wr^2\left(\mathrm{S}-\mathfrak{g}\frac{\sigma}{a}\right)=w\tau\sin 2\epsilon . xy \quad . \quad . \quad . \quad . \quad . \quad . \quad . \quad (20)$$

It was also shown in the paper on "Tides" that the solution of Sir W. THOMSON'S problem of the state of internal strain of an elastic sphere, devoid of gravitation, as distorted by a bodily force, of which the potential is expressible as a solid harmonic function of the second degree, is identical in form with the solution of the parallel problem for a viscous spheroid.

That solution is as follows :—

$$\alpha=\frac{1}{19v}\left[(4a^2-\tfrac{21}{10}r^2)\frac{d\mathrm{W}}{dx}-\tfrac{2}{5}r^7\frac{d}{dx}\left(\frac{\mathrm{W}}{r^5}\right)\right]^*$$

with symmetrical expressions for β and γ.

Since $\frac{d}{dx}\left(\frac{\mathrm{W}}{r^5}\right)=\frac{1}{r^5}\frac{d\mathrm{W}}{dx}-\frac{5x}{r^7}\mathrm{W}$, the solution may be written

$$\alpha=\frac{1}{38v}\left[(8a^2-5r^2)\frac{d\mathrm{W}}{dx}+4x\mathrm{W}\right],\ \beta=\&\mathrm{c.},\ \gamma=\&\mathrm{c.}$$

Then substituting for W from (20) we have

$$\left.\begin{aligned}\alpha&=\frac{w\tau}{38v}\sin 2\epsilon\left[(8a^2-5r^2)y+4x^2y\right]\\ \beta&=\frac{w\tau}{38v}\sin 2\epsilon\left[(8a^2-5r^2)x+4xy^2\right]\\ \gamma&=\frac{w\tau}{38v}\sin 2\epsilon\ 4xyz\end{aligned}\right\} \quad . \quad . \quad . \quad . \quad . \quad . \quad (21)$$

Putting $\mathrm{K}=\frac{w\tau}{19v}\sin 2\epsilon$, we have

$$\left.\begin{aligned}&\frac{d\alpha}{dx}=-\mathrm{K}xy,\ \frac{d\alpha}{dy}=\tfrac{1}{2}\mathrm{K}[8a^2-(x^2+15y^2+5z^2)],\ \frac{d\alpha}{dz}=-5\mathrm{K}yz\\ &\frac{d\beta}{dx}=\tfrac{1}{2}\mathrm{K}[8a^2-(15x^2+y^2+5z^2)],\ \frac{d\beta}{dy}=-\mathrm{K}xy,\ \frac{d\beta}{dz}=-5\mathrm{K}xz\\ &\frac{d\gamma}{dx}=2\mathrm{K}yz,\qquad \frac{d\gamma}{dy}=2\mathrm{K}zx,\qquad \frac{d\gamma}{dz}=2\mathrm{K}xy\end{aligned}\right\} \quad . \quad . \quad . \quad (22)$$

And

$$\frac{d\beta}{dz}+\frac{d\gamma}{dy}=-3\mathrm{K}zx,\ \frac{d\gamma}{dx}+\frac{d\alpha}{dz}=-3\mathrm{K}yz,\ \frac{d\alpha}{dy}+\frac{d\beta}{dx}=\mathrm{K}[8(a^2-x^2-y^2)-5z^2] \quad . \quad (23)$$

* See THOMSON and TAIT'S 'Nat. Phil.,' § 834, or "Tides," Section 3.

Now if P, Q, R, S, T, U be the stresses across three mutually rectangular planes at x, y, z, estimated in the usual way, then the work done per unit time on a unit of volume situated at x, y, z is

$$\mathrm{P}\frac{d\alpha}{dx}+\mathrm{Q}\frac{d\beta}{dy}+\mathrm{R}\frac{d\gamma}{dz}+\mathrm{S}\left(\frac{d\beta}{dz}+\frac{d\gamma}{dy}\right)+\mathrm{T}\left(\frac{d\gamma}{dx}+\frac{d\alpha}{dz}\right)+\mathrm{U}\left(\frac{d\alpha}{dy}+\frac{d\beta}{dx}\right)^*$$

But $\mathrm{P}=-p+2v\frac{d\alpha}{dx}$, $\mathrm{S}=v\left(\frac{d\beta}{dz}+\frac{d\gamma}{dy}\right)$, and Q, R, T, U have symmetrical forms. Therefore, substituting in the expression for the work $\left(\text{which will be called } \frac{d\mathrm{E}}{dt}\right)$, and remembering that

$$\frac{d\alpha}{dx}+\frac{d\beta}{dy}+\frac{d\gamma}{dz}=0,$$

we have

$$\frac{1}{v}\frac{d\mathrm{E}}{dt}=2\left\{\left(\frac{d\alpha}{dx}\right)^2+\left(\frac{d\beta}{dy}\right)^2+\left(\frac{d\gamma}{dz}\right)^2\right\}+\left(\frac{d\beta}{dz}+\frac{d\gamma}{dy}\right)^2+\left(\frac{d\gamma}{dx}+\frac{d\alpha}{dz}\right)^2+\left(\frac{d\alpha}{dy}+\frac{d\beta}{dx}\right)^2$$

Now from (22)

$$\frac{2}{\mathrm{K}^2}\left[\left(\frac{d\alpha}{dx}\right)^2+\left(\frac{d\beta}{dy}\right)^2+\left(\frac{d\gamma}{dz}\right)^2\right]=12x^2y^2=\tfrac{3}{2}r^4\sin^4\theta[1-\cos 4(\phi-\omega t)] \quad . \quad . \quad (24)$$

and from (23)

$$\frac{1}{\mathrm{K}^2}\left[\left(\frac{d\beta}{dz}+\frac{d\gamma}{dy}\right)^2+\left(\frac{d\gamma}{dx}+\frac{d\alpha}{dz}\right)^2+\left(\frac{d\alpha}{dy}+\frac{d\beta}{dx}\right)^2\right]=9z^2(x^2+y^2)+[8(a^2-x^2-y^2)-5z^2]^2$$

$$=9r^4\sin^2\theta\cos^2\theta+(8a^2-5r^2-3r^2\sin^2\theta)^2 \quad . \quad . \quad . \quad (25)$$

Adding (24) and (25) and rearranging the terms

$$\frac{1}{\mathrm{K}^2v}\frac{d\mathrm{E}}{dt}=-\tfrac{3}{2}r^4\sin^4\theta\cos 4(\phi-\omega t)+(8a^2-5r^2)^2-\tfrac{3}{2}r^2\sin^2\theta[32a^2-r^2(26+\sin^2\theta)].$$

The first of these terms is periodic, going through its cycle of changes in six lunar hours, and therefore the average rate of work, or the average rate of heat generation, is given by

$$\frac{d\mathrm{E}}{dt}=\frac{1}{v}\left(\frac{w\tau}{19}\sin 2\epsilon\right)^2[(8a^2-5r^2)^2-\tfrac{3}{2}r^2\sin^2\theta\{32a^2-r^2(26+\sin^2\theta)\}] \quad . \quad . \quad (26)$$

It will now be well to show that this formula leads to the same results as those given in the paper on "Precession."

In order to find the whole heat generated per unit time throughout the sphere, we must find the integral $\iiint\frac{d\mathrm{E}}{dt}r^2\sin\theta dr d\theta d\phi$, from $r=a$ to 0, $\theta=\pi$ to 0, $\phi=2\pi$ to 0.

* THOMSON and TAIT, 'Nat. Phil.,' § 670.

In a later investigation we shall require a transformation of the expression for $\frac{dE}{dt}$, and as it will here facilitate the integration, it will be more convenient to effect the transformation now.

If Q_2, Q_4 be the zonal harmonics of the second and fourth order,

$$\cos^2\theta = \tfrac{2}{3}Q_2 + \tfrac{1}{3},$$
$$\cos^4\theta = \tfrac{8}{35}Q_4 + \tfrac{4}{7}Q_2 + \tfrac{1}{5}.*$$

Now

$$(8a^2-5r^2)^2 - \tfrac{3}{2}r^2\sin^2\theta[32a^2-(26+\sin^2\theta)r^2]$$
$$=(8a^2-5r^2)^2 - r^2[48a^2-\tfrac{81}{2}r^2-\tfrac{3}{2}(32a^2-28r^2)\cos^2\theta-\tfrac{3}{2}r^2\cos^4\theta]$$
$$=\tfrac{1}{5}\{320a^4-560a^2r^2+259r^4\}-\tfrac{2}{7}(112a^2-95r^2)r^2Q_2+\tfrac{12}{35}r^4Q_4 \quad . \quad . \quad . \quad (27)$$

The last transformation being found by substituting for $\cos^2\theta$ and $\cos^4\theta$ in terms of Q_2 and Q_4, and rearranging the terms.

The integrals of Q_2 and Q_4 vanish when taken all round the sphere, and

$$\iiint\tfrac{1}{5}(320a^4-560a^2r^2+259r^4)r^2\sin\theta\, dr\, d\theta\, d\phi = \frac{4\pi a^7}{5}\{\tfrac{320}{3}-\tfrac{560}{5}+\tfrac{259}{7}\} = \frac{Ca^2}{w}\times\tfrac{5}{2}\times 19,$$

where C is the earth's moment of inertia, and therefore equal to $\frac{8}{15}\pi w a^5$.

Hence we have

$$\iiint\frac{dE}{dt}r^2\sin\theta\, dr\, d\theta\, d\phi = \frac{w}{v}\left(\frac{\tau}{19}\sin 2\epsilon\right)^2 Ca^2.\tfrac{5}{2}\times 19 = \frac{5wa^2}{38v}(\tau\sin 2\epsilon)^2 C.$$

But $\tan 2\epsilon = \frac{19v\omega}{gaw} = 2.\frac{19v\omega}{5\mathfrak{g}wa^2}$, so that $\frac{5wa^2}{38v} = \frac{\omega}{\mathfrak{g}}\cot 2\epsilon$.

And the whole work done on the sphere per unit time is $\frac{1}{2}\frac{\tau^2}{\mathfrak{g}}\sin 4\epsilon . C\omega$.

Now, as shown in the first part (equation 5), if $\mathfrak{N}$ be the tidal frictional couple $\frac{\mathfrak{N}}{C} = \frac{1}{2}\frac{\tau^2}{\mathfrak{g}}\sin 4\epsilon$.

Therefore the work done on the sphere per unit time is $\mathfrak{N}\omega$.

It is worth mentioning, in passing, that if the integral be taken from $\frac{1}{2}a$ to 0, we find that ·32 of the whole heat is generated within the central eighth of the volume; and by taking the integral from $\frac{7}{8}a$ to a, we find that one-tenth of the whole heat is generated within 500 miles of the surface.

It remains to show the identity of this remarkably simple result, for the whole work done on the sphere, with that used in the paper on "Precession." It was there shown

* TODHUNTER'S 'Functions of LAPLACE,' &c., p. 13; or any other work on the subject.

(Section 16) that if n be the earth's rotation, r the moon's distance at any time, ν the ratio of the earth's mass to the moon's, then the whole energy both potential and kinetic of the moon-earth system is

$$\tfrac{1}{2}C\left(n^2-\frac{5g}{2\nu}\cdot\frac{1}{r}\right).$$

Now c being the moon's distance initially, since the lunar orbit is supposed to be circular,

$$\Omega_0^2c^3=ga^2\frac{1+\nu}{\nu}.$$

Also

$$\frac{c}{r}=\left(\frac{\Omega}{\Omega_0}\right)^{\frac{2}{3}}=\frac{1}{\xi^2}.$$

Therefore

$$\tfrac{2}{5}\frac{c\nu}{g}=\tfrac{2}{5}\left\{\left(\frac{a}{g}\right)^2\nu^2(1+\nu)\right\}^{\frac{1}{3}}\Omega_0^{-\frac{2}{3}}=s\Omega_0^{-\frac{2}{3}},$$

according to the notation of the paper on "Precession."

In that paper I also put $\dfrac{1}{\mu}=sn_0\Omega_0^{\frac{1}{3}}$.

Therefore $\dfrac{5g}{2\nu}\cdot\dfrac{1}{r}=\dfrac{\mu n_0\Omega_0}{\xi^2}$.

And the whole energy of the system is $\tfrac{1}{2}C\left(n^2-\dfrac{\mu n_0\Omega_0}{\xi^2}\right)$.

Therefore the rate of loss of energy is $-C\left(n\dfrac{dn}{dt}+\dfrac{\mu n_0}{\xi^3}\Omega_0\dfrac{d\xi}{dt}\right)$.

But $\dfrac{dn}{dt}=-\dfrac{\mathfrak{N}}{C}$, and as shown in the first part (19), $\mu n_0\dfrac{d\xi}{dt}=\dfrac{\mathfrak{N}}{C}$, also $\dfrac{\Omega_0}{\xi^3}=\Omega$.

Therefore the rate of loss of energy is $\mathfrak{N}(n-\Omega)$ or $\mathfrak{N}\omega$, which expression agrees with that obtained above. The two methods therefore lead to the same result.

I will now return to the investigation in hand.

The average throughout the earth of the rate of loss of energy is $\mathfrak{N}\omega\div\tfrac{4}{3}\pi a^3$, which quantity will be called H. Then

$$H=\frac{3}{4\pi a^3}\mathfrak{N}\omega=\frac{w}{M}\cdot\tfrac{2}{5}Ma^2.\tfrac{1}{2}\frac{\tau^2}{\mathfrak{g}}\sin 4\epsilon.\omega=\tfrac{1}{5}wa^2.\frac{\tau^2}{\mathfrak{g}}\sin 4\epsilon.\omega.$$

Now

$$\frac{1}{v}\left(\frac{w\tau}{19}\sin 2\epsilon\right)^2a^4=\frac{2\omega}{5\mathfrak{g}}\cot 2\epsilon.\frac{\tau^2}{19}\sin^2 2\epsilon.wa^2=\tfrac{1}{19}\cdot\tfrac{1}{5}wa^2.\frac{\tau^2}{\mathfrak{g}}\sin 4\epsilon.\omega=\frac{H}{19}.$$

Hence (26) may be written

$$\frac{dE}{dt}=\frac{H}{19}\left[\left\{8-5\left(\frac{r}{a}\right)^2\right\}^2-\tfrac{3}{2}\left(\frac{r}{a}\right)^2\sin^2\theta\left\{32-(26+\sin^2\theta)\left(\frac{r}{a}\right)^2\right\}\right]\quad . \quad . \quad . \quad (28)$$

This expression gives the rate of generation of heat at any point in terms of the average rate, and if we equate it to a constant we get the equation to the family of surfaces of equal heat-generation.

We may observe that the heat generated at the centre is $3\frac{7}{19}$ times the average, at the pole $\frac{1}{2\frac{1}{11}}$ of the average, and at the equator $\frac{1}{12\frac{2}{3}}$ of the average.

The accompanying figure exhibits the curves of equal heat-generation; the dotted line shows that of $\frac{1}{4}$ of the average, and the others those of $\frac{1}{2}$, 1, $1\frac{1}{2}$, 2, $2\frac{1}{2}$, and 3 times the average. It is thus obvious from inspection of the figure that by far the largest part of the heat is generated in the central regions.

Fig. 2.

The next point to consider is the effect which the generation of heat will have on underground temperature, and how far it may modify the investigation of the secular cooling of the earth.

It has already been shown* that the total amount of heat which might be generated is very large, and my impression was that it might, to a great extent, explain the increase of temperature underground, until a conversation with Sir W. THOMSON led me to undertake the following calculations:—

We will first calculate in what length of time the earth is losing by cooling an amount of energy equal to its present kinetic energy of rotation.

The earth's conductivity may be taken as about ·004 according to the results given in EVERETT's illustrations of the centimeter-gram-second system of units, and the temperature gradient at the surface as 1° C. in $27\frac{1}{2}$ meters, which is the same as 1° Fahr. in 50 feet—the rate used by Sir W. THOMSON in his paper on the cooling of the earth.†

This temperature gradient is $\frac{4}{11 \times 10^3}$ degrees C. per centimeter, and since there are 31,557,000 seconds in a year, therefore in centimeter-gram-second units,

* "Precession," Section 15, Table IV., and Section 16.

† THOMSON and TAIT's 'Nat. Phil.,' Appendix D.

The heat lost by earth per annum $\Big\} = \text{earth's surface in square centimeters} \times \frac{4}{10^3} \times \frac{4}{11 \times 10^3} \times 3{\cdot}1557 \times 10^7$

$= \text{earth's surface} \times 45{\cdot}9$ (centimeter-gram-second heat units).

Now if J be JOULE'S equivalent

Earth's kinetic energy of rotation in heat units $\Big\} = \frac{1}{2}\frac{Cn_0^2}{gJ} = \frac{Ma}{J}(\tfrac{2}{5})^2\left(\frac{5}{4}\frac{n_0^2 a}{g}\right)$, where $C = \frac{2}{5}Ma^2$

$= \text{earth's surface} \times \frac{wa^2}{3J}(\tfrac{2}{5})^2 e_0$, where $e_0 = \frac{5}{4}\frac{n_0^2 a}{g} = \frac{1}{232}$

$= \text{earth's surface} \times \frac{(5{\cdot}5) \times (6{\cdot}37)^2 \times 10^{16} \times (\cdot 4)^2}{3 \times 4{\cdot}34 \times 10^4 \times 232}$, for $a = 6{\cdot}37 \times 10^8$ centimeters. $J = 4{\cdot}34 \times 10^4$ gram centim. and $w = 5\frac{1}{2}$.

$= \text{earth's surface} \times 1{\cdot}2 \times 10^{10}$ nearly.

Therefore at the present rate of loss the earth is losing energy by cooling equivalent to its kinetic energy of rotation in $\frac{1{\cdot}2 \times 10^{10}}{45{\cdot}9} = 262$ million years.

If we had taken the earth as heterogeneous and $C = \frac{1}{3}Ma^2$ we should have found 218 million years.

We will next find how much energy is lost to the moon-earth system in the series of changes investigated in the paper on "Precession."

In that paper (Section 16) it was shown that the whole energy of the system is $\frac{1}{5}Ma^2\left(n^2 - \frac{5g}{2\nu}\frac{1}{r}\right)$, where ν is earth ÷ moon, r moon's distance, n earth's diurnal rotation.

Hence the loss of energy $= \frac{1}{5}Ma^2 n_0^2\left[\left(\frac{n}{n_0}\right)^2 - 1 - \frac{5g}{2\nu n_0^2}\left(\frac{1}{r} - \frac{1}{r_0}\right)\right]$, while n passes from n to n_0, and r from r to r_0.

Now $\frac{5g}{2\nu n_0^2} = \frac{25}{8\nu}\left(\frac{4g}{5n_0^2 a}\right)a = \frac{100 \times 232}{32 \times 82}a = 8{\cdot}84a$, taking $\nu = 82$, and $\frac{4g}{5n_0^2 a} = 232$.

If D be the length of the day, $\frac{n}{n_0} = \frac{D_0}{D}$; and if Π be the moon's distance in earth's radii, then

$$\text{loss of energy} = \left[\left(\frac{D_0}{D}\right)^2 - 1 - 8{\cdot}84\left(\frac{1}{\Pi} - \frac{1}{\Pi_0}\right)\right] \times \text{earth's present } k.e. \text{ of rotation.}$$

But in the paper on "Precession" we showed the system passing from a day of 5 hours 40 minutes,* and a lunar distance of 2·547 earth's radii, to a day of 24 hours, and a lunar distance of 60·4 earth's radii.

* A recalculation in the paper on "Precession" gave 5 hours 36 minutes, but I have not thought it worth while to alter this calculation.

Now $24 \div 5\frac{2}{3} = 4{\cdot}23$, and $(2{\cdot}547)^{-1} - (60{\cdot}4)^{-1} = {\cdot}376$.

Therefore the loss of energy $= [(4{\cdot}23)^2 - 1 - {\cdot}376 \times 8{\cdot}84] \times$ earth's present *k.e.*
$= 13{\cdot}57 \times$ earth's present *k.e.* of rotation.

Hence the whole heat, generated in the earth from first to last, gives a supply of heat, at the present rate of loss, for $13{\cdot}6 \times 262$ million years, or 3,560 million years.

This amount of heat is certainly prodigious, and I found it hard to believe that it should not largely affect the underground temperature. But Sir W. THOMSON pointed out to me that the distribution of its generation would probably be such as not materially to affect the temperature gradient at the earth's surface; this remarkable prevision on his part has been confirmed by the results of the following problem, which I thought might be taken to roughly represent the state of the case.

Conceive an infinite slab of rock of thickness $2a$ (or 8,000 miles) being part of an infinite mass of rock; suppose that in a unit of volume, distant x from the medial plane, there is generated, per unit time, a quantity of heat equal to $\mathfrak{h}[320a^4 - 560a^2x^2 + 259x^4]$; suppose that initially the slab and the whole mass of rock have a uniform temperature V; let the heat begin to be generated according to the above law, and suppose that the two faces of the slab are for ever maintained at the constant temperature V; then it is required to find the distribution of temperature within the slab after any time.

This problem roughly represents the true problem to be considered, because if we replace x by the radius vector r, we have the average distribution of internal heat-generation due to friction; also the maintenance of the faces of the slab at a constant temperature represents the rapid cooling of the earth's surface, as explained by Sir W. THOMSON in his investigation.

Let ϑ be temperature, γ thermal capacity, k conductivity; then the equation of heat-flow is

$$\gamma \frac{d\vartheta}{dt} = k\frac{d^2\vartheta}{dx^2} + \mathfrak{h}[320a^4 - 560a^2x^2 + 259x^4].$$

Let $320\frac{\mathfrak{h}}{k} = 2\text{L}$, $560\frac{\mathfrak{h}}{k} = 12\text{M}$, $259\frac{\mathfrak{h}}{k} = 30\text{N}$, and let the thermometric conductivity $\kappa = \frac{k}{\gamma}$. Then

$$\frac{d\vartheta}{dt} = \kappa\frac{d^2}{dx^2}[\vartheta + \text{L}a^4x^2 - \text{M}a^2x^4 + \text{N}x^6 - \text{R}].$$

Let the constant $\text{R} = (\text{L} - \text{M} + \text{N})a^6$, and put

$$\begin{aligned}\psi &= \vartheta + \text{L}a^4x^2 - \text{M}a^2x^4 + \text{N}x^6 - \text{R}\\ &= \vartheta - \text{L}a^4(a^2 - x^2) + \text{M}a^2(a^4 - x^4) - \text{N}(a^6 - x^6).\end{aligned}$$

Then when $x = \pm a$, $\psi = \vartheta$.

Since L, M, N, R are constants as regards the time,

$$\frac{d\psi}{dt}=\kappa\frac{d^2\psi}{dx^2}.$$

$\psi=V-\Sigma Pe^{-\kappa q^2 t}\cos qx$ is obviously a solution of this equation.

Now we wish to make $\vartheta=V$, when $x=\pm a$, for all values of t; since $\psi=\vartheta$ when $x=\pm a$, this condition is clearly satisfied by making $q=(2i+1)\frac{\pi}{2a}$.

Hence the solution may be written,

$$\vartheta=V-[La^4x^2-Ma^2x^4+Nx^6-R]-\sum_0^\infty P_{2i+1}e^{-\kappa t\left[\frac{(2i+1)\pi}{2a}\right]^2}\cos(2i+1)\frac{\pi x}{2a} \quad . \quad (29)$$

and it satisfies all the conditions except that, initially, when $t=0$, the temperature everywhere should be V. This last condition is satisfied if

$$\sum_0^\infty P_{2i+1}\cos(2i+1)\frac{\pi x}{2a}=R-La^4x^2+Ma^2x^4-Nx^6$$

for all values between $x=\pm a$.

The expression on the right must therefore be expanded by FOURIER'S Theorem; but we need only consider the range from $x=a$ to 0, because the rest, from $x=0$ to $-a$, will follow of its own accord.

Let $\chi=\frac{\pi x}{2a}$; let ϖ be written for $\frac{\pi}{2}$; let $M'=\frac{M}{\varpi^2}$, $N'=\frac{N}{\varpi^4}$ and $R'=R\frac{\varpi^2}{a^6}$.

Then

$$R-La^4x^2+Ma^2x^4-Nx^6=\frac{a^6}{\varpi^2}[R'-L\chi^2+M'\chi^4-N'\chi^6],$$

and this has to be equal to $\sum_0^\infty P_{2i+1}\cos(2i+1)\chi$ from $\chi=\frac{\pi}{2}$ to 0.

Since

$$\int_0^{\frac{\pi}{2}}\cos(2i+1)\chi\cos(2j+1)\chi d\chi=0 \text{ unless } j=i,$$

and

$$\int_0^{\frac{\pi}{2}}\cos^2(2i+1)\chi d\chi=\tfrac{1}{4}\pi=\tfrac{1}{2}\varpi,$$

Therefore

$$\tfrac{1}{2}\varpi P_{2i+1}=\frac{a^6}{\varpi^2}\int_0^{\frac{\pi}{2}}[R'-L\chi^2+M'\chi^4-N'\chi^6]\cos(2i+1)\chi d\chi.$$

Now

$$\int_0^{\frac{\pi}{2}} \chi^{2j} \cos(2i+1)\chi d\chi = \frac{1}{2i+1}\left[\chi^{2j} \sin(2i+1)\chi + \frac{\frac{d\chi^{2j}}{d\chi}}{2i+1}\cos(2i+1)\chi - \frac{\frac{d^2\chi^{2j}}{d\chi^2}}{(2i+1)^2}\sin(2i+1)\chi - \frac{\frac{d^3\chi^{2j}}{d\chi^3}}{(2i+1)^3}\cos(2i+1)\chi + \&c.\right]_0^{\frac{\pi}{2}}$$

$$= \frac{(-)^i}{2i+1}\left[1 - \frac{\frac{d^2}{d\varpi^2}}{(2i+1)^2} + \frac{\frac{d^4}{d\varpi^4}}{(2i+1)^4} - \&c.\right]\varpi^{2j}.$$

If therefore $f(\chi)$ be a function of χ involving only even powers of χ,

$$\int_0^{\frac{\pi}{2}} f(\chi)\cos(2i+1)\chi d\chi = \frac{(-)^i}{(2i+1)}\left[1+\left(\frac{1}{2i+1}\frac{d}{d\varpi}\right)^2\right]^{-1} f(\varpi).$$

This theorem will make the calculation of the coefficients very easy, for we have at once

$$\frac{\varpi^3}{2a^6}P_{2i+1} = \frac{(-)^i}{2i+1}\Big\{R' - L\varpi^2 + M'\varpi^4 - N'\varpi^6$$
$$- \frac{1}{(2i+1)^2}[-2L + 4.3M'\varpi^2 - 6.5N'\varpi^4]$$
$$+ \frac{1}{(2i+1)^4}[4.3.2.1.M' - 6.5.4.3N'\varpi^2]$$
$$- \frac{1}{(2i+1)^6}[-6.5.4.3.2.1N']\Big\}.$$

Substituting for R′, L, M′, N′ their values in terms of $\frac{\mathfrak{h}}{k}$ we find

$$P_{2i+1} = \frac{(-)^i 2a^6}{(2i+1)^3\varpi^3}\frac{\mathfrak{h}}{k}\left[19 - \frac{1988}{(2i+1)^2\varpi^2} + \frac{6216}{(2i+1)^4\varpi^4}\right].$$

Then putting for ϖ its value, viz.: ½ of 3·14159, and putting i successively equal to 0, 1, 2, it will be found that

$$P_1 = \frac{\mathfrak{h}a^6}{k}(120{\cdot}907),\ P_3 = \frac{\mathfrak{h}a^6}{k}(1{\cdot}107),\ P_5 = -\frac{\mathfrak{h}a^6}{k}({\cdot}048).$$

So that the Fourier expansion is

$$120{\cdot}907\cos\frac{\pi x}{2a} + 1{\cdot}107\cos\frac{3\pi x}{2a} - {\cdot}048\cos\frac{5\pi x}{2a},$$

which will be found to differ by not so much as one per cent. from the function

$$\tfrac{320}{2}\left(1-\left(\frac{x}{a}\right)^2\right)-\tfrac{560}{12}\left(1-\left(\frac{x}{a}\right)^4\right)+\tfrac{259}{30}\left(1-\left(\frac{x}{a}\right)^6\right),$$

to which it should be equal.

Then by substitution in (29) we have as the complete solution of the problem satisfying all the conditions

$$\vartheta=\mathrm{V}+\frac{\mathfrak{h}a^6}{k}\left\{\left(1-e^{-\kappa\left(\frac{\pi}{2a}\right)^2t}\right)120\cdot907\cos\frac{\pi x}{2a}\right.$$
$$\left.+\left(1-e^{-\kappa\left(\frac{3\pi}{2a}\right)^2t}\right)1\cdot107\cos\frac{3\pi x}{2a}-\left(1-e^{-\kappa\left(\frac{5\pi}{2a}\right)^2t}\right)\cdot048\cos\frac{5\pi x}{2a}\right\}.$$

The only quantity, which it is of interest to determine, is the temperature gradient at the surface, which is equal to $-\frac{d\vartheta}{dx}$ when $x=\pm a$.

Now when $x=\pm a$,

$$\frac{d\vartheta}{dx}=\frac{\mathfrak{h}a^5}{k}\frac{\pi}{2}\left\{120\cdot907\left(1-e^{-\kappa\left(\frac{\pi}{2a}\right)^2t}\right)-3\cdot321\left(1-e^{-\kappa\left(\frac{3\pi}{2a}\right)^2t}\right)-\cdot240\left(1-e^{-\kappa\left(\frac{5\pi}{2a}\right)^2t}\right)\right\}.$$

Then if t be not so large but that $\kappa\left(\frac{5\pi}{2a}\right)^2t$ is a small fraction, we have approximately

$$-\frac{d\vartheta}{dx}=\frac{\mathfrak{h}a^5}{k}\frac{\pi}{2}\kappa\left(\frac{\pi}{2a}\right)^2t\left\{120\cdot907-9\times3\cdot321-25\times\cdot240\right\};$$

and since $\frac{\kappa}{k}=\frac{1}{\gamma}$

$$-\frac{d\vartheta}{dx}=\left(\frac{\pi a}{2}\right)^3\frac{\mathfrak{h}}{\gamma}t\times(85).$$

This formula will give the temperature gradient at the surface when a proper value is assigned to $\mathfrak{h}$, and if t be not taken too large.

With respect to the value of t, Sir W. THOMSON took $\kappa=400$ in British units, the year being the unit of time; and $a=21\times10^6$ feet.

Hence

$$\kappa\left(\frac{\pi}{2a}\right)^2=4\times10^2\left(\frac{1\cdot5}{2\cdot1\times10^7}\right)^2=\frac{2}{10^{12}}\text{ nearly,}$$

and $\kappa\left(\frac{5\pi}{2a}\right)^2=\frac{5}{10^{11}}$; if therefore t be 10^9 years, this fraction is $\frac{1}{20}$. Therefore the solution given above will hold provided the time t does not exceed 1,000 million years.

We next have to consider what is the proper value to assign to $\mathfrak{h}$.

By (27) and (28) it appears that $\mathfrak{h}a^4$ is $\frac{1}{5\times19}$ of the average heat generated throughout the whole earth, which we called H. Suppose that p times the present kinetic energy of the earth's rotation is destroyed by friction in a time T, and suppose the generation of heat to be uniform in time, then the average heat generated throughout the whole earth per unit time is

$$\frac{p}{g\mathrm{JT}}\cdot\tfrac{1}{5}\mathrm{M}a^2n_0{}^2\div\text{earth's volume.}$$

Therefore

$$\mathrm{H}=\frac{p}{5\mathrm{JT}}\cdot\frac{wa^2n_0{}^2}{g}=\tfrac{4}{25}\,\frac{p}{\mathrm{JT}}\,wae_0.$$

Where e_0 is the ellipticity of figure of the homogeneous earth and is equal to $\frac{5}{4}\,\frac{n_0{}^2a}{g}$, which I take as equal to $\frac{1}{232}$.

Hence

$$\mathfrak{h}a^4=\tfrac{16}{9500}\,\frac{p}{\mathrm{JT}}\,wae_0,$$

and

$$-\frac{d\vartheta}{dx}=\frac{16\times 85}{9500}\left(\frac{\pi}{2}\right)^3\frac{w}{\gamma}\,\frac{pe_0}{\mathrm{J}}\,\frac{t}{\mathrm{T}}.$$

But $\gamma=sw$, where s is specific heat.

Therefore

$$-\frac{d\vartheta}{dx}=\frac{170\pi^3}{9500}\,\frac{pe_0}{s}\,\frac{1}{\mathrm{J}}\,\frac{t}{\mathrm{T}}.$$

The dimensions of J are those of work (in gravitation units) per mass and per scale of temperature, that is to say, length per scale of temperature; p, e_0, and s have no dimensions, and therefore this expression is of proper dimensions.

Now suppose the solution to run for the whole time embraced by the changes considered in "Precession," then $t=\mathrm{T}$, and as we have shown $p=13{\cdot}57$. Suppose the specific heat to be that of iron, viz.: $\frac{1}{9}$. Then if we take $\mathrm{J}=772$, so that the result will be given in degrees Fahrenheit per foot, we have

$$-\frac{d\vartheta}{dx}=\frac{17\pi^3}{950}\times\frac{13{\cdot}57\times 9}{232\times 772}$$

$$=\frac{1}{2650}.$$

That is to say, at the end of the changes the temperature gradient would be 1° Fahr. per 2,650 feet, provided the whole operation did not take more than 1,000 million years.

It might, however, be thought that if the tidal friction were to operate very slowly, so that the whole series of changes from the day of 5 hours 36 minutes to that of 24 hours occupied much more than 1,000 million years, then the large amount of heat which is generated deep down would have time to leak out, so that finally the temperature gradient would be steeper than that just found. But this is not the case.

Consider only the first, and by far the most important, term of the expression for the temperature gradient. It has the form $\mathfrak{h}\,(1-e^{-pT})$, when $t=T$ at the end of the series of changes. Now $\mathfrak{h}$ varies as T^{-1}, and $\frac{1-e^{-pT}}{pT}$ has its maximum value unity when $T=0$. Hence, however slowly the tidal friction operates, the temperature gradient can never be greater than if the heat were all generated instantaneously; but the temperature gradient at the end of the changes is not sensibly less than it would be if all the heat were generated instantaneously, provided the series of changes do not occupy more than 1,000 million years.

III. *The forced oscillations of viscous, fluid, and elastic spheroids.*

In investigating the tides of a viscous spheroid, the effects of inertia were neglected, and it was shown that the neglect could not have an important influence on the results.* I shall here obtain an approximate solution of the problem including the effects of inertia; that solution will easily lead to a parallel one for the case of an elastic sphere, and a comparison with the forced oscillations of a fluid spheroid will prove instructive as to the nature of the approximation.

If W be the potential of the impressed forces, estimated per unit volume of the viscous body, then (with the same notation as before) the equations of flow are

$$\left.\begin{aligned}&-\frac{dp}{dx}+\upsilon\nabla^2\alpha+\frac{dW}{dx}-w\left(\frac{d\alpha}{dt}+\alpha\frac{d\alpha}{dx}+\beta\frac{d\alpha}{dy}+\gamma\frac{d\alpha}{dz}\right)=0\\&-\frac{dp}{dy}+\&c.=0\quad-\frac{dp}{dz}+\&c.\ =0\\&\frac{d\alpha}{dx}+\frac{d\beta}{dy}+\frac{d\gamma}{dz}=0\end{aligned}\right\} \quad . \qquad (30)$$

The terms $-w\left(\frac{d\alpha}{dt}+\&c.\right)$ are those due to inertia, which were neglected in the paper on "Tides."

It will be supposed that the tidal motion is steady, and that W consists of a series of solid harmonics each multiplied by a simple time harmonic, also that W includes not only the potential of the external tide-generating body, but also the effective potential due to gravitation, as explained in the first part of this paper.

* "Tides," Section 10.

The tidal disturbance is supposed to be sufficiently slow to enable us to obtain a first approximation by the neglect of the inertia terms.

In proceeding to the second approximation, the inertia terms depending on the squares and products of the velocities, that is to say, $w\left(\alpha\frac{d\alpha}{dx}+\beta\frac{d\alpha}{dy}+\gamma\frac{d\alpha}{dz}\right)$, may be neglected compared with $w\frac{d\alpha}{dt}$. A typical case will be considered in which $W=Y\cos(vt+\epsilon)$, where Y is a solid harmonic of the ith degree, and the ϵ will be omitted throughout the analysis for brevity. Then if we write $I=2(i+1)^2+1$, the first approximation, when the inertia terms are neglected, is

$$\alpha=\frac{1}{Iv}\left\{\left[\frac{i(i+2)}{2(i-1)}a^2-\frac{(i+1)(2i+3)}{2(2i+1)}r^2\right]\frac{dY}{dx}-\frac{i}{2i+1}r^{2i+3}\frac{d}{dx}(r^{-2i-1}Y)\right\}\cos vt^{*} \quad . \quad . \quad (31)$$

Hence for the second approximation we must put

$$-w\frac{d\alpha}{dt}=\frac{wv}{Iv}\left\{\cdots\right\}\sin vt.$$

And the equations to be solved are

$$\left.\begin{array}{l}-\frac{dp}{dx}+v\nabla^2\alpha+\frac{dY'}{dx}\cos vt+\frac{wv}{Iv}\left\{\left[\frac{i(i+2)}{2(i-1)}a^2-\frac{(i+1)(2i+3)}{2(2i+1)}r^2\right]\frac{dY}{dx}\right.\\ \qquad\qquad\qquad\qquad\left.-\frac{i}{2i+1}r^{2i+3}\frac{d}{dx}(r^{-2i-1}Y)\right\}\sin vt=0\\ -\frac{dp}{dy}+\text{\&c.}=0, \qquad -\frac{dp}{dz}+\text{\&c.}=0\end{array}\right\} \quad . \quad (32)$$

These equations are to be satisfied throughout a sphere subject to no surface stress. It will be observed that in the term due directly to the impressed forces, we write Y′ instead of Y; this is because the effective potential due to gravitation will be different in the second approximation from what it was in the first, on account of the different form which must now be attributed to the tidal protuberance.

The problem is now reduced to one strictly analogous to that solved in the paper on "Tides;" for we may suppose that the terms introduced by $w\frac{d\alpha}{dt}$ &c., are components of bodily force acting on the viscous spheroid, and that inertia is neglected.

The equations being linear, we consider the effects of the several terms separately, and indicate the partial values of α, β, γ, p by suffixes and accents.

First, then, we have

$$-\frac{dp_0}{dx}+v\nabla^2\alpha_0+\frac{dY'}{dx}\cos vt=0, \text{ \&c., \&c.}$$

* "Tides," Section 3, equation (8), or THOMSON and TAIT, 'Nat. Phil.,' § 834 (8).

The solution of this has the same form as in the first approximation, viz.: equation (31), with α_0 written for α, and Y' for Y.

We shall have occasion hereafter to use the velocity of flow resolved along the radius vector, which may be called ρ. Then

$$\rho_0=\alpha_0\frac{x}{r}+\beta_0\frac{y}{r}+\gamma_0\frac{z}{r}.$$

Hence

$$\rho_0=\frac{1}{\mathrm{I}v}\left\{\frac{i^2(i+2)a^2-i(i^2-1)r^2}{2(i-1)}\right\}\frac{Y'}{r}\cos vt \quad . \quad . \quad . \quad . \quad . \quad (33)$$

Then observing that $Y'\div r^i$ is independent of r, we have as the surface value

$$\rho_0=\frac{a^{i+1}}{\mathrm{I}v}\,\frac{i(2i+1)}{2(i-1)}\,\frac{Y'}{r^i}\cos vt \quad . \quad (34)$$

Secondly,

$$-\frac{dp_0'}{dx}+v\nabla^2\alpha'_0+\frac{wva^2}{\mathrm{I}v}\,\frac{i(i+2)}{2(i-1)}\,\frac{dY}{dx}\sin vt=0, \text{ \&c., \&c.} \quad . \quad . \quad . \quad . \quad . \quad (35)$$

This, again, may clearly be solved in the same way, and we have

$$\alpha'_0=\frac{wva^2}{\mathrm{I}^2v^2}\cdot\frac{i(i+2)}{2(i-1)}\left\{\left[\frac{i(i+2)}{2(i-1)}a^2-\frac{(i+1)(2i+3)}{2(2i+1)}r^2\right]\frac{dY}{dx}-\frac{i}{2i+1}r^{2i+3}\frac{d}{dx}(Yr^{-2i-1})\right\}\sin vt \quad (36)$$

and

$$\rho_0'=\frac{wva^2}{\mathrm{I}^2v^2}\cdot\frac{i(i+2)}{2(i-1)}\left\{\frac{i^2(i+2)a^2-i(i^2-1)r^2}{2(i-1)}\right\}\frac{Y}{r}\sin vt \quad (37)$$

and its surface value is

$$\rho_0'=wva^{i+3}\cdot\frac{i^2(i+2)(2i+1)}{[2\,\mathrm{I}v(i-1)]^2}\,\frac{Y}{r^i}\sin vt \quad . \quad . \quad . \quad . \quad . \quad . \quad . \quad . \quad (38)$$

Thirdly, let

$$U=\frac{wv}{\mathrm{I}v}\,\frac{Y}{2(2i+1)}\sin vt \quad . \quad . \quad . \quad . \quad . \quad . \quad . \quad . \quad . \quad . \quad (39)$$

So that U is a solid harmonic of the ith degree multiplied by a simple time harmonic. Then the rest of the terms to be satisfied are given in the following equations:—

$$\left.\begin{aligned}&-\frac{dp}{dx}+v\nabla^2\alpha-\left[(i+1)(2i+3)r^2\frac{dU}{dx}+2ir^{2i+3}\frac{d}{dx}(Ur^{-2i-1})\right]=0\\&-\frac{dp}{dy}+\text{ \&c. }=0,\qquad -\frac{dp}{dz}+\text{ \&c. }=0\end{aligned}\right\} \quad . \quad . \quad . \quad . \quad (40)$$

These equations have to be satisfied throughout a sphere subject to no surface stresses. The procedure will be exactly that explained in Part I., viz.: put $\alpha=\alpha'+\alpha_{,}$

$\beta=\beta'+\beta_{,}$, $\gamma=\gamma'+\gamma_{,}$, $p=p'+p_{,}$, and find α', β', γ', p' any functions which satisfy the equations (40) throughout the sphere.

Differentiate the three equations (40) by x, y, z respectively and add them together, and notice that

$$(i+1)(2i+3)\left\{\frac{d}{dx}\left(r^2\frac{dU}{dx}\right)+\frac{d}{dy}\left(\ \right)+\frac{d}{dz}\left(\ \right)\right\}+2i\left\{\frac{d}{dx}\left(r^{2i+3}\frac{d}{dx}Ur^{-2i-1}\right)+\frac{d}{dy}\left(\ \right)+\frac{d}{dz}\left(\ \right)\right\}=0,$$

and that

$$\frac{d\alpha'}{dx}+\frac{d\beta'}{dy}+\frac{d\gamma'}{dz}=0\,;$$

then we have $\nabla^2p'=0$, of which $p'=0$ is a solution.

Now if V_n be a solid harmonic of degree n,

$$\nabla^2r^mV_n=m(2n+m+1)r^{m-2}V_n$$

Hence

$$\left.\begin{aligned}r^2\frac{dU}{dx}&=\nabla^2\frac{r^4}{4(2i+3)}\frac{dU}{dx}\\ r^{2i+3}\frac{d}{dx}(Ur^{-2i-1})&=\nabla^2\frac{r^{2i+5}}{2(2i+5)}\frac{d}{dx}(Ur^{-2i-1})\end{aligned}\right\}\quad\ldots\ldots\quad(41)$$

Substituting from (41) in the equations of motion (40), and putting $p'=0$, our equations become

$$\left.\begin{aligned}&\nabla^2\left\{v\alpha'-\frac{(i+1)}{4}r^4\frac{dU}{dx}-\frac{i}{2i+5}r^{2i+5}\frac{d}{dx}(Ur^{-2i-1})\right\}=0\\ &\nabla^2\{v\beta'-\&c.\}=0,\ \nabla^2\{v\gamma'-\&c.\}=0\end{aligned}\right\}\quad\ldots\ldots\quad(42)$$

of which a solution is obviously

$$\left.\begin{aligned}&\alpha'=\frac{1}{v}\left\{\frac{i+1}{4}r^4\frac{dU}{dx}+\frac{i}{2i+5}r^{2i+5}\frac{d}{dx}(Ur^{-2i-1})\right\}\\ &\beta'=\&c.,\ \gamma'=\&c.\end{aligned}\right\}\quad\ldots\ldots\quad(43)$$

It may easily be shown that these values satisfy the equation of continuity, and thus together with $p'=0$ they are the required values of α', β', γ', p', which satisfy the equations throughout the sphere.

The next step is to find the surface stresses to which these values give rise. The formulas (13) of Part I. are applicable

$$\begin{aligned}v\zeta'&=v(\alpha'x+\beta'y+\gamma'z)\\ &=\frac{i(i+1)}{4}r^4U-\frac{i(i+1)}{2i+5}r^4U=\frac{i(i+1)(2i+1)}{4(2i+5)}r^4U.\end{aligned}$$

Then remembering that

$$xU=\frac{1}{2i+1}\left\{r^2\frac{dU}{dx}-r^{2i+3}\frac{d}{dx}(r^{-2i-1}U)\right\},$$

We have

$$v\frac{d\zeta'}{dx}=\frac{i(i+1)(2i+1)}{4(2i+5)}\left\{r^4\frac{dU}{dx}+\frac{4r^2}{2i+1}r^2\frac{dU}{dx}-\frac{4r^2}{2i+1}r^{2i+3}\frac{d}{dx}(r^{-2i-1}U)\right\}$$

$$=\frac{i(i+1)}{4(2i+5)}\left\{(2i+5)r^4\frac{dU}{dx}-4r^{2i+5}\frac{d}{dx}(r^{-2i-1}U)\right\} \quad . \quad . \quad . \quad . \quad . \quad . \quad (44)$$

Again, by the properties of homogeneous functions,

$$v\left(r\frac{d}{dr}-1\right)\alpha'=v\left(x\frac{d}{dx}+y\frac{d}{dy}+z\frac{d}{dz}\right)\alpha'-v\alpha'$$

$$=\frac{(i+1)(i+2)}{4}r^4\frac{dU}{dx}+\frac{i(i+2)}{2i+5}r^{2i+5}\frac{d}{dx}(r^{-2i-1}U) \quad . \quad . \quad . \quad . \quad (45)$$

Also $p'=0$.

Then adding (44) and (45) together, we have for the component of stress parallel to the axis of x across any of the concentric spherical surfaces,

$$Fr=-p'x+v\left[\left(r\frac{d}{dr}-1\right)\alpha'+\frac{d\zeta'}{dx}\right] \text{ by (13), Part I.}$$

$$=\frac{(i+1)^2}{2}r^4\frac{dU}{dx}+\frac{i}{2i+5}r^{2i+5}\frac{d}{dx}(r^{-2i-1}U) \text{ by (44) and (45).}$$

And at the surface of the sphere, where $r=a$,

$$F=\frac{(i+1)^2}{2}a^{i+2}\left[r^{-i+1}\frac{dU}{dx}\right]+\frac{i}{2i+5}a^{i+2}\left[r^{i+2}\frac{d}{dx}(r^{-2i-1}U)\right] \quad . \quad . \quad . \quad . \quad (46)$$

The quantities in square brackets are independent of r, and are surface harmonics of orders $i-1$ and $i+1$ respectively.

Let

$$F=-A_{i-1}-A_{i+1},$$

Where

$$A_{i-1}=-\frac{(i+1)^2}{2}a^{i+2}\left[r^{-i+1}\frac{dU}{dx}\right], \quad A_{i+1}=-\frac{i}{2i+5}a^{i+2}\left[r^{i+2}\frac{d}{dx}(r^{-2i-1}U)\right] \quad . \quad (47)$$

Also let the other two components G and H of the surface stress due to α', β', γ, p' be given by

$$G=-B_{i-1}-B_{i+1}, \quad H=-C_{i-1}-C_{i+1} \quad . \quad . \quad . \quad . \quad . \quad . \quad . \quad (47)$$

Then by symmetry it is clear that the B's and C's only differ from the A's in having y and z in place of x.

We now have got in (43) values of α', β', γ', which satisfy the equations (40) throughout the sphere, together with the surface stresses in (47) to which they correspond. Thus (43) would be the solution of the problem, if the surface of the sphere were subject to the surface stresses (47). It only remains to find $\alpha_{,}, \beta_{,}, \gamma_{,}$, to satisfy the equations

$$-\frac{dp_{,}}{dx}+v\nabla^{2}\alpha_{,}=0,\ -\frac{dp_{,}}{dy}+\&c.=0,\ -\frac{dp_{,}}{dz}+\&c.=0 \quad . \quad . \quad . \quad . \quad (48)$$

throughout the sphere, which is not under the influence of bodily force, but is subject to surface stresses of which $A_{i-1}+A_{i+1}$, $B_{i-1}+B_{i+1}$, $C_{i-1}+C_{i+1}$ are the components.

The sum of the solution of these equations and of the solutions (43) will clearly be the complete solution; for (43) satisfies the condition as to the bodily force in (40), and the two sets of surface actions will annul one another, leaving no surface action.

For the required solutions of (48), Sir W. THOMSON's solution given in (15) and (16) of Part I. is at once applicable.

We have first to find the auxiliary functions Ψ_{i-2}, Φ_i corresponding to A_{i-1}, B_{i-1}, C_{i-1}, and Ψ_i, Φ_{i+2} corresponding to A_{i+1}, B_{i+1}, C_{i+1}. It is easy to show that

$$\Psi_{i-2}=0,\ \Phi_{i+2}=0,$$

and

$$\Psi_i=-a^{i+2}\frac{i}{2i+5}\left[\frac{d}{dx}\left\{r^{2i+3}\frac{d}{dx}(r^{-2i-1}U)\right\}+\frac{d}{dy}\left\{\quad\right\}+\frac{d}{dz}\left\{\quad\right\}\right]$$

$$=a^{i+2}\frac{i(i+1)(2i+3)}{2i+5}U$$

$$\Phi_i=-r^{2i+1}a^{i+2}\frac{(i+1)^2}{2}\left[\frac{d}{dx}\left(r^{-2i+1}\frac{dU}{dx}\right)+\frac{d}{dy}\left(\quad\right)+\frac{d}{dz}\left(\quad\right)\right]$$

$$=a^{i+2}\frac{i(i+1)^2(2i-1)}{2}U.$$

We have next to substitute these values of the auxiliary functions in THOMSON's solution (15), Part I. It will be simpler to perform the substitutions piece-meal, and to indicate the various parts which go to make up the complete value of $\alpha_{,}$ by accents to that symbol.

First. For the terms in $\alpha_{,}$ depending on A_{i-1}, Ψ_{i-2}, Φ_i, we have

$$\alpha_{,}'=\frac{1}{va^{i-2}}\left\{\frac{1}{2(i-2)(i-1)(2i-1)}\frac{d\Phi_i}{dx}+\frac{1}{i-2}A_{i-1}r^{i-1}\right\}$$

$$=\frac{a^4}{v}\left\{\frac{i(i+1)^2}{4(i-1)(i-2)}\frac{dU}{dx}-\frac{(i+1)^2}{2(i-2)}\frac{dU}{dx}\right\}$$

$$=-\frac{a^4}{v}\frac{(i+1)^2}{4(i-1)}\frac{dU}{dx} \quad . \quad . \quad . \quad . \quad . \quad . \quad . \quad . \quad . \quad . \quad (49)$$

(Note that $i-2$ divides out, so that the solution is still applicable when $i=2$).

Second. In finding the terms dependent on A_{i+1}, Ψ_i, Φ_{i+2} it will be better to subdivide the process further.

$$\text{(i)}\quad \alpha_{,}''=\frac{1}{va^i}\frac{1}{2\,\text{I}}(a^2-r^2)\frac{d\Psi_i}{dx}$$

$$=\frac{a^2}{v}\frac{i(i+1)(2i+3)}{2\,\text{I}(2i+5)}(a^2-r^2)\frac{d\text{U}}{dx} \quad \ldots \quad (50)$$

$$\text{(ii)}\quad \alpha_{,}'''=\frac{1}{va^i}\left\{\frac{i+3}{\text{I}i(2i+3)}r^{2i+3}\frac{d}{dx}(r^{-2i-1}\Psi_i)+\frac{1}{i}\text{A}_{i+1}r^{i+1}\right\}$$

$$=\frac{a^2}{v}\left\{\frac{(i+1)(i+3)}{\text{I}(2i+5)}r^{2i+3}\frac{d}{dx}(r^{-2i-1}\text{U})-\frac{1}{2i+5}r^{2i+3}\frac{d}{dx}(r^{-2i-1}\text{U})\right\}.$$

Then since

$$(i+3)(i+1)-\text{I}=i^2+4i+3-2i^2-4i-3=-i^2,$$

therefore

$$\alpha_{,}'''=-\frac{a^2}{v}\frac{i^2}{\text{I}(2i+5)}r^{2i+3}\frac{d}{dx}(r^{-2i-1}\text{U}) \quad \ldots \quad (51)$$

This completes the solution for $\alpha_{,}$.

Collecting results from (49), (50), and (51), we have

$$\alpha_{,}=\alpha_{,}'+\alpha_{,}''+\alpha_{,}'''$$

$$=-\frac{a^2}{v}\left\{\frac{(i+1)^2}{4(i-1)}a^2\frac{d\text{U}}{dx}-\frac{i(i+1)(2i+3)}{2\,\text{I}(2i+5)}(a^2-r^2)\frac{d\text{U}}{dx}+\frac{i^2}{\text{I}(2i+5)}r^{2i+3}\frac{d}{dx}(r^{-2i-1}\text{U})\right\} \quad . \quad (52)$$

Then collecting results, the complete value of α as the solution of the second approximation is

$$\alpha=\alpha_0+\alpha_0'+\alpha'+\alpha_{,}.$$

So that it is only necessary to collect the results of equations (31), (with Y′ written for Y), (36), (43), and (52), and to substitute for U its value from (39) in order to obtain the solution required. The values of β and γ may then at once be written down by symmetry. The expressions are naturally very long, and I shall not write them down in the general case.

The radial velocity ρ is however an important expression, because it alone is necessary to enable us to obtain the second approximation to the form of the spheroid, and accordingly I will give it.

It may be collected from (33), (37), and by forming ρ' and $\rho_{,}$ from (43) and (52).

I find then after some rather tedious analysis, which I did in order to verify my solution, that as far as concerns the inertia terms alone

$$\rho=\frac{wv}{v^2}\frac{Y}{r}\sin vt\{\mathfrak{A}r^4-\mathfrak{B}a^2r^2+\mathfrak{C}a^4\},$$

Where

$$\mathfrak{A}=\frac{i(i+1)}{2.4(2i+5)\text{I}},\quad \mathfrak{B}=\frac{i^2(i+1)(2i^2+10i+9)}{4(i-1)(2i+5)\text{I}^2}, \text{ and}$$

$$\mathfrak{C}=\left(\frac{i}{2\text{I}}\right)^2\left[i\left(\frac{i+2}{i-1}\right)^2+\frac{(i+1)(2i+3)}{(2i+1)(2i+5)}\right]-\frac{i(i+1)^2}{2.4(i-1)(2i+1)\text{I}}.$$

If $\mathfrak{C}$ be reduced to the form of a single fraction, I think it probable that the numerator would be divisible by $2i+1$, but I do not think that the quotient would divide into factors, and therefore I leave it as it stands.

In the case where $i=2$ this formula becomes

$$\rho=\frac{wv}{v^2}\frac{Y}{r}\sin vt\frac{1}{2^2.3.19^2}\{19r^4-148a^2r^2+287a^4\},$$

which agrees (as will appear presently) with the same result obtained in a different way.

I shall now go on to the special case where $i=2$, which will be required in the tidal problem.

From (39) we have

$$\text{U}=\frac{wv}{v}\cdot\frac{1}{2.5.19}\text{Y}\sin vt.$$

From (36)

$$\alpha_0'=\frac{wva^2}{v^2}\cdot\frac{4}{19^2}\left[\left(4a^2-\frac{3.7}{2.5}r^2\right)\frac{d\text{Y}}{dx}-\frac{2}{5}r^7\frac{d}{dx}(\text{Y}r^{-5})\right]\sin vt.$$

From (43)

$$\alpha'=\frac{wv}{v^2}\cdot\frac{1}{2^3.3.5.19}\left[9r^4\frac{d\text{Y}}{dx}+\frac{2.4}{3}r^9\frac{d}{dx}(\text{Y}r^{-5})\right]\sin vt.$$

From (52)

$$\alpha_,=-\frac{wva^2}{v^2}\cdot\frac{1}{2^3.3.5.19^2}\left[(5.97a^2+7.4r^2)\frac{d\text{Y}}{dx}-\frac{4.4}{3}r^7\frac{d}{dx}r^{-5}\text{Y})\right].$$

Adding these expressions together, and adding α_0, we get

$$\alpha=\alpha_0+\frac{wv}{v^2}\cdot\frac{1}{2^3.3.5.19^2}\left[(5.287a^4-37.4.7a^2r^2+9.19r^4)\frac{d\text{Y}}{dx}\right.$$

$$\left.-\tfrac{8}{3}(2.37a^2-19r^2)r^7\frac{d}{dx}(\text{Y}r^{-5})\right]\sin vt \quad . \quad (53)$$

and symmetrical expressions for β and γ.

In order to obtain the radial flow we multiply α by $\frac{x}{r}$, β by $\frac{y}{r}$, γ by $\frac{z}{r}$, and add, and find

$$\rho=\rho_0+\frac{wv}{v^2}\cdot\frac{1}{2^2.3.19^2}(287a^4-4.37a^2r^2+19r^4)\frac{\text{Y}}{r}\sin(vt+\epsilon) \quad . \quad (54)$$

the ϵ which was omitted in the trigonometrical term being now replaced.

The surface value of ρ when $r=a$ is

$$\rho=\rho_0+\frac{wva^5}{v^2}\frac{79}{2.3.19^2}\frac{Y}{r^2}\sin(vt+\epsilon) \tag{55}$$

where ρ_0 is given by (34).

If we write $-\frac{1}{2}\pi-\epsilon$ for ϵ we see that a term Y sin $(vt-\epsilon)$ in the effective disturbing potential will give us

$$\rho=\rho_0-\frac{wva^5}{v^2}\frac{79}{2.3.19^2}\frac{Y}{r^2}\cos(vt-\epsilon)\ . \tag{56}$$

Now suppose wr^2S cos vt to be an external disturbing potential per unit volume of the earth, not including the effective potential due to gravitation, and let $r=a+\sigma_{\prime}$ be the first approximation to the form of the tidal spheroid. Then by the theory of tides as previously developed (see equation (15), Section 5, "Tides")

$$\frac{\sigma_{\prime}}{a}=\frac{S}{\mathfrak{g}}\cos\epsilon\cos(vt-\epsilon),\ \text{where}\ \tan\epsilon=\frac{19vv}{2gaw}$$

Then when the sphere is deemed free of gravitation the effective disturbing potential is $wr^2\left(S\cos vt-\mathfrak{g}\frac{\sigma_{\prime}}{a}\right)$; this is equal to $-wr^2\sin\epsilon\ S\sin(vt-\epsilon)$.

Then in proceeding to a second approximation we must put in equation (56) $Y=-wr^2\sin\epsilon\ S$.

Thus we get from (56), at the surface where $r=a$,

$$\rho=\rho_0+\frac{w^2va^5}{v^2}\cdot\frac{79}{2.3.19^2}\sin\epsilon\ S\cos(vt-\epsilon) \quad . \quad . \tag{57}$$

To find ρ_0 we must put $r=a+\sigma$ as the equation to the second approximation.

Then ρ_0 is the surface radial velocity due directly to the external disturbing potential wr^2S cos vt and to the effective gravitation potential. The sum of these two gives an effective potential $wr^2\left(S\cos vt-\mathfrak{g}\frac{\sigma}{a}\right)$, which is the Y′ cos vt of (34).

Then ρ_0 is found by writing this expression in place of Y′ cos vt in equation (34), and we have

$$\rho_0=\frac{5wa^3}{19v}\left(S\cos vt-\mathfrak{g}\frac{\sigma}{a}\right).$$

Substituting in (57) we have

$$\rho=\frac{5wa^3}{19v}\left(S\cos vt-\mathfrak{g}\frac{\sigma}{a}+\frac{5wva^2}{19v}\frac{79}{2.3.5^3}\sin\epsilon\ S\cos(vt-\epsilon)\right)\quad . \quad . \quad . \quad . \tag{58}$$

Then since $\tan\epsilon=\frac{19vv}{2gaw}$, therefore $\frac{5wa^2}{19v}=\frac{v}{\mathfrak{g}}\cot\epsilon$, and (58) becomes

$$\rho=a\frac{v}{\mathfrak{g}}\cot\epsilon\left(\mathrm{S}\cos vt-\mathfrak{g}\frac{\sigma}{a}+\frac{79}{150}\frac{v^2}{\mathfrak{g}}\cos\epsilon\,\mathrm{S}\cos(vt-\epsilon)\right).$$

But the radial surface velocity is equal to $\frac{d\sigma}{dt}$, and therefore $\frac{d\sigma}{dt}=\rho$, so that

$$\frac{d\sigma}{dt}+v\cot\epsilon.\sigma=a\frac{v}{\mathfrak{g}}\cot\epsilon\left(\mathrm{S}\cos vt+\frac{79}{150}\frac{v^2}{\mathfrak{g}}\cos\epsilon\,\mathrm{S}\cos(vt-\epsilon)\right)\quad . \quad . \quad . \quad (59)$$

Then if we divide σ into two parts, σ', σ'', to satisfy the two terms on the right respectively, we have

$$\frac{\sigma'}{a}=\cos\epsilon\cdot\frac{\mathrm{S}}{\mathfrak{g}}\cos(vt-\epsilon),$$

which is the first approximation over again, and

$$\frac{\sigma''}{a}=\cos\epsilon\cdot\frac{\mathrm{S}}{\mathfrak{g}}\cdot\frac{79}{150}\frac{v^2}{\mathfrak{g}}\cos\epsilon\cos(vt-2\epsilon).$$

Therefore

$$\frac{\sigma}{a}=\cos\epsilon\cdot\frac{\mathrm{S}}{\mathfrak{g}}\left\{\cos(vt-\epsilon)+\frac{79}{150}\frac{v^2}{\mathfrak{g}}\cos\epsilon\cos(vt-2\epsilon)\right\}\quad . \quad . \quad . \quad . \quad (60)$$

This gives the second approximation to the form of the tidal spheroid. We see that the inertia generates a second small tide which lags twice as much as the primary one.

Although this expression is more nearly correct than subsequent ones, it will be well to group both these tides together and to obtain a single expression for σ.

Let

$$\tan\chi=\frac{\frac{79}{150}\frac{v^2}{\mathfrak{g}}\sin\epsilon\cos\epsilon}{1+\frac{79}{150}\frac{v^2}{\mathfrak{g}}\cos^2\epsilon},$$

Then

$$\frac{\sigma}{a}=\frac{\mathrm{S}}{\mathfrak{g}}\frac{\cos\epsilon}{\cos\chi}\left(1+\tfrac{79}{150}\frac{v^2}{\mathfrak{g}}\cos^2\epsilon\right)\cos(vt-\epsilon-\chi)\quad . \quad (61)$$

This shows that the tide lags by $(\epsilon+\chi)$, and is in height $\frac{\cos\epsilon}{\cos\chi}\left(1+\tfrac{79}{150}\frac{v^2}{\mathfrak{g}}\cos^2\epsilon\right)$ of the equilibrium tide of a perfectly fluid spheroid.

By the method employed it is postulated that $\tfrac{79}{150}\frac{v^2}{\mathfrak{g}}$ is a small fraction, because the effects of inertia are supposed to be small. Hence χ must be a small angle, and there will not be much error in putting

$$\chi=\tfrac{79}{150}\frac{v^2}{\mathfrak{g}}\sin\epsilon\cos\epsilon,\text{ and }\sec\chi=1.$$

Then we have for the lag of the tide $\left(\epsilon+\tfrac{79}{150}\frac{v^2}{\mathfrak{g}}\sin\epsilon\cos\epsilon\right)$, and for its height $\cos\epsilon\left(1+\tfrac{79}{150}\frac{v^2}{\mathfrak{g}}\cos^2\epsilon\right)$.

Let η be the lag, then

$$\eta=\epsilon+\tfrac{79}{150}\frac{v^2}{\mathfrak{g}}\sin\epsilon\cos\epsilon,$$

whence

$$\epsilon=\eta-\tfrac{79}{150}\frac{v^2}{\mathfrak{g}}\sin\eta\cos\eta\text{ very nearly.}$$

Also

$$\cos\epsilon=\cos\eta\left(1+\tfrac{79}{150}\frac{v^2}{\mathfrak{g}}\sin^2\eta\right),$$

and

$$\cos\epsilon\left(1+\tfrac{79}{150}\frac{v^2}{\mathfrak{g}}\cos^2\epsilon\right)=\cos\eta\left(1+\tfrac{79}{150}\frac{v^2}{\mathfrak{g}}\right).$$

Hence (61) becomes

$$\left.\begin{aligned}\frac{\sigma}{a}&=\frac{S}{\mathfrak{g}}\cos\eta\left(1+\tfrac{79}{150}\frac{v^2}{\mathfrak{g}}\right)\cos(vt-\eta),\\ \text{where}\quad \eta-\tfrac{79}{150}\frac{v^2}{\mathfrak{g}}\sin\eta\cos\eta&=\text{arc tan}\left(\frac{19\upsilon v}{2\mathfrak{g}aw}\right)\end{aligned}\right\}\quad . \quad (62)$$

This is probably the simplest form in which the result of the second approximation may be stated.

From it we see that with a given lag, the height of tide is a little greater than in the theories used in the two previous papers; and that for a given frequency of tide the lag is a little greater than was supposed.

The whole investigation of the precession of the viscous spheroid was based on the approximate theory of tides, when inertia is neglected. It will be well, therefore, to examine how far the present results will modify the conclusions there arrived at. It would, however, occupy too much space to recapitulate the methods employed, and therefore the following discussion will only be intelligible, when read in conjunction with that paper.

The couples on the earth, caused by the attraction of the disturbing bodies on the tidal protuberance, were found to be expressible by the sum of a number of terms, each of which corresponded to one of the constituent simple harmonic tides. Each such term involved two factors, one of which was the height of the tide, and the other the sine of the lag. Now if ϵ be the lag and v the speed of the tide, it was found in the first approximation that $\tan\epsilon = 19vv \div 2gaw$, and that the height of tide was proportional to $\cos\epsilon$; hence each term had a factor $\sin 2\epsilon$.

But from the present investigation it appears that, with the same value of ϵ, the height of tide is really proportional to $\cos\epsilon\left(1+\frac{79}{150}\frac{v^2}{\mathfrak{g}}\cos^2\epsilon\right)$; whilst the lag is $\epsilon+\frac{79}{150}\frac{v^2}{\mathfrak{g}}\sin\epsilon\cos\epsilon$, so that its sine is $\left(1+\frac{79}{150}\frac{v^2}{\mathfrak{g}}\cos^2\epsilon\right)\sin\epsilon$.

Hence in place of $\sin 2\epsilon$, we ought to have put $\sin 2\epsilon\left(1+\frac{79}{150}\frac{v^2}{\mathfrak{g}}\cos^2\epsilon\right)^2$, or $\sin 2\epsilon\left(1+\frac{79}{75}\frac{v^2}{\mathfrak{g}}\cos^2\epsilon\right)$.

Thus every term in the expressions for $\frac{di}{dt}, \frac{dN}{dt}, \frac{d\xi}{dt}$ should be augmented, each in a proportion depending on the speed and lag of the tide from which it takes its origin.

* In the paper on "Precession," two numerical integrations were given of the differential equations for the secular changes in the variables; in the first of these, in Section 15, the viscosity was not supposed to be small, and was constant, in the second, in Section 17, it was merely supposed that the alteration of phase of each tide was small, and the viscosity was left indeterminate. It is not proposed to determine directly the correction to the first solution.

The correcting factor for the expression $\sin 2\epsilon$ is greatest when ϵ is small, because $\cos^2\epsilon$ may then be replaced in it by unity; hence the correction in the second integration will necessarily be larger than in the first, and a superior limit to the correction to the first integration may be found.

We have tides of the seven speeds $2(n-\Omega)$, $2n$, $2(n+\Omega)$, $n-2\Omega$, n, $n+2\Omega$, 2Ω; hence if the viscosity be small, the correcting factors for the expressions $\sin 4\epsilon_1$, $\sin 4\epsilon$, $\sin 4\epsilon_2$, $\sin 2\epsilon'_1$, $\sin 2\epsilon'$, $\sin 2\epsilon'_2$, $\sin 4\epsilon''$ are respectively $1+\frac{79}{75}\frac{1}{\mathfrak{g}}$ multiplied by the squares of the above seven speeds.

Then if $\lambda=\frac{\Omega}{n}$, the seven factors may be written

$$\left.\begin{array}{l} 1+\frac{316}{75}\frac{n^2}{\mathfrak{g}} \text{ multiplied by } (1-\lambda)^2,\ 1,\ (1+\lambda)^2, \text{ for semi-diurnal terms} \\ 1+\frac{79}{75}\frac{n^2}{\mathfrak{g}} \text{ multiplied by } (1-2\lambda)^2,\ 1,\ (1+2\lambda)^2, \text{ for diurnal terms} \\ \text{and } 1+\frac{316}{75}\frac{n^2}{\mathfrak{g}}\lambda^2, \text{ for the fortnightly term} \end{array}\right\} \quad . \quad (63)$$

* The following method of correcting the work of the paper on "Precession" has been rewritten, and was inserted on the 17th May, 1879.

Also we have the equations

$$\left.\begin{array}{l}\frac{\sin 4\epsilon_1}{\sin 4\epsilon}=1-\lambda, \quad \frac{\sin 4\epsilon}{\sin 4\epsilon}=1, \quad \frac{\sin 4\epsilon_2}{\sin 4\epsilon}=1+\lambda \\ \frac{\sin 2\epsilon'_1}{\sin 4\epsilon}=\tfrac{1}{2}(1-2\lambda), \quad \frac{\sin 2\epsilon'}{\sin 4\epsilon}=\tfrac{1}{2}, \quad \frac{\sin 2\epsilon'_2}{\sin 4\epsilon}=\tfrac{1}{2}(1+2\lambda), \quad \frac{\sin 4\epsilon''}{\sin 4\epsilon}=\lambda\end{array}\right\} \quad (64)$$

Now we shall obtain a sufficiently accurate result, if the corrections be only applied to those terms in the differential equations which do not involve powers of q (or $\sin \frac{1}{2}i$), higher than the first. Then for the purpose of correction the differential equations to be corrected are by (77), (78), and (79) of Section 17 of "Precession," viz. :—

$$\left.\begin{array}{l}\frac{di_{m^2}}{dt}=\frac{1}{N}\frac{\tau^2}{\mathfrak{g}n_0}\left[\tfrac{1}{2}p^7q \sin 4\epsilon_1+\tfrac{1}{2}p^5q(p^2+3q^2)\sin 2\epsilon'_1-\tfrac{1}{2}pq(p^2-q^2)^3 \sin 2\epsilon'\right] \\ -\frac{dN_{m^2}}{dt}=\tfrac{1}{2}\frac{\tau^2}{\mathfrak{g}n_0}p^8 \sin 4\epsilon_1=\mu\frac{d\xi}{dt}\end{array}\right\} \quad (65)$$

As we are treating the obliquity as small, we may put

$$\tfrac{1}{2}p^7q=\tfrac{1}{2}p^5q(p^2+3q^2)=\tfrac{1}{2}pq(p^2-q^2)^3=\tfrac{1}{4}PQ \text{ and } p^8=P,$$

when $P=\cos i$, $Q=\sin i$.

Then for the purpose of correction, the terms depending on the moon's influence are

$$\left.\begin{array}{l}\frac{di_{m^2}}{dt}=\frac{1}{N}\frac{\tau^2}{\mathfrak{g}n_0}\tfrac{1}{4}PQ\{\sin 4\epsilon_1+\sin 2\epsilon'_1-\sin 2\epsilon'\} \\ -\frac{dN_{m^2}}{dt}=\tfrac{1}{2}\frac{\tau^2}{\mathfrak{g}n_0}P \sin 4\epsilon_1=\mu\frac{d\xi}{dt}\end{array}\right\} \quad (66)$$

And by symmetry (or by (81) "Precession") we have for the solar terms

$$\frac{di_{m_i^2}}{dt}=\frac{1}{N}\frac{\tau_i^2}{\mathfrak{g}n_0}\tfrac{1}{4}PQ \sin 4\epsilon, \qquad -\frac{dN_{m_i^2}}{dt}=\tfrac{1}{2}\frac{\tau_i^2}{\mathfrak{g}n_0}P \sin 4\epsilon \quad (67)$$

For the terms depending on the joint action of the sun and moon we have, by (82) and (33) "Precession," when the obliquity is treated as small,

$$\left.\begin{array}{l}\frac{di_{mm_i}}{dt}=-\frac{1}{N}\frac{\tau\tau_i}{\mathfrak{g}n_0}\tfrac{1}{2}PQ \sin 2\epsilon' \\ \frac{dN_{mm_i}}{dt}=0\end{array}\right\} \quad (68)$$

Then if we multiply each of the sines by its appropriate factor given in (63), and substitute from (64) for each of them in terms of $\sin 4\epsilon$, and collect the results from (66), (67), and (68), and express by the symbol δ the corrections to be introduced for the effects of inertia, we have

$$\delta\frac{di}{dt}=\frac{1}{N}\frac{\sin 4\epsilon}{\mathfrak{g}n_0}\tfrac{1}{4}PQ\,.\,\tfrac{79}{75}\frac{n^2}{\mathfrak{g}}\left[\left\{4(1-\lambda)^3+\tfrac{1}{2}(1-2\lambda)^3-\tfrac{1}{2}\right\}\tau^2+4\tau_{\prime}^2-\tau\tau_{\prime}\right]$$

$$-\delta\frac{dN}{dt}=\tfrac{1}{2}\frac{\sin 4\epsilon}{\mathfrak{g}n_0}P\,.\,\tfrac{316}{75}\frac{n^2}{\mathfrak{g}}\left[\tau^2(1-\lambda)^3+\tau_{\prime}^2\right]$$

$$\mu\delta\frac{d\xi}{dt}=\tfrac{1}{2}\frac{\sin 4\epsilon}{\mathfrak{g}n_0}P\,.\,\tfrac{316}{75}\frac{n^2}{\mathfrak{g}}\tau^2(1-\lambda)^3.$$

Now $4(1-\lambda)^3+\frac{1}{2}(1-2\lambda)^3-\frac{1}{2}=(1-2\lambda)(4-7\lambda+4\lambda^2)$. Therefore if we add these corrections to the full expressions for $\frac{di}{dt}$, $\frac{dN}{dt}$ (in which I put $1-\frac{1}{2}Q^2=P$) and $\mu\frac{d\xi}{dt}$, given in (83) "Precession," and write $\mathrm{K}=\frac{316}{75}\frac{n^2}{\mathfrak{g}}$ for brevity, we have

$$\left.\begin{aligned}
\frac{di}{dt}&=\frac{1}{N}\frac{\sin 4\epsilon}{\mathfrak{g}n_0}\tfrac{1}{4}PQ\left\{\tau^2\left(1-\frac{2\lambda}{P}\right)+\tau_{\prime}^2-\tau\tau_{\prime}+\mathrm{K}\left[(1-2\lambda)(1-\tfrac{7}{4}\lambda+\lambda^2)\tau^2+\tau_{\prime}^2-\tfrac{1}{4}\tau\tau_{\prime})\right]\right\}\\
-\frac{dN}{dt}&=\tfrac{1}{2}\frac{\sin 4\epsilon}{\mathfrak{g}n_0}P\left\{\tau^2(1-\lambda)+\tau_{\prime}^2+\tfrac{1}{2}\tau\tau_{\prime}\frac{Q^2}{P}+\mathrm{K}\left[(1-\lambda)^3\tau^2+\tau_{\prime}^2\right]\right\}\\
\mu\frac{d\xi}{dt}&=\tfrac{1}{2}\frac{\tau^2}{\mathfrak{g}n_0}\sin 4\epsilon P\left[1-\frac{\lambda}{P}+\mathrm{K}(1-\lambda)^3\right]
\end{aligned}\right\}\quad(69)$$

The last of these equations may be written approximately

$$\frac{dt}{\mu d\xi}=\left[\tfrac{1}{2}\frac{\tau^2}{\mathfrak{g}n_0}\sin 4\epsilon\, P\left(1-\frac{\lambda}{P}\right)\right]^{-1}\left[1-\mathrm{K}(1-\lambda)^2\right]. \quad . \quad . \quad . \quad . \quad . \quad (70)$$

Then if we multiply the two former of equations (69) by (70), and notice that, when P is taken as unity,

$$(1-2\lambda)(1-\tfrac{7}{4}\lambda+\lambda^2)-\left(1-\frac{2\lambda}{P}\right)(1-\lambda)^2=\tfrac{1}{4}\lambda(1-2\lambda),$$

and that

$$1-(1-\lambda)^2=\lambda(2-\lambda)\text{ and }-\tfrac{1}{4}+(1-\lambda)^2=\tfrac{1}{4}(1-2\lambda)(3-2\lambda).$$

we have

$$\left.\begin{aligned}&\frac{d}{\mu d\xi}\log\tan^2\frac{i}{2}\\&\quad=\frac{1-\frac{2\lambda}{P}+\left(\frac{\tau_{\prime}}{\tau}\right)^2-\left(\frac{\tau_{\prime}}{\tau}\right)+\mathrm{K}\left[\frac{1}{4}\lambda(1-2\lambda)+\lambda(2-\lambda)\left(\frac{\tau_{\prime}}{\tau}\right)^2+\frac{1}{4}(1-2\lambda)(3-2\lambda)\left(\frac{\tau_{\prime}}{\tau}\right)\right]}{N\left(1-\frac{\lambda}{P}\right)}\\&-\frac{dN}{\mu d\xi}=\frac{1-\lambda+\left(\frac{\tau_{\prime}}{\tau}\right)^2+\frac{1}{2}\frac{Q^2}{P}\left(\frac{\tau_{\prime}}{\tau}\right)+\mathrm{K}\lambda(2-\lambda)\left(\frac{\tau_{\prime}}{\tau}\right)^2}{1-\frac{\lambda}{P}}\end{aligned}\right\}\quad(71)$$

If K be put equal to zero, we have the equations (84) which were the subject of integration in Section 17 "Precession."

Since K, λ, and $\tau_{\prime}^2\div\tau^2$ are all small, the correction to the second equation is obviously insignificant, and we may take the term in K in the numerator of the first equation as being equal to $\frac{1}{4}\mathrm{K}(1-2\lambda)(3-2\lambda)(\tau_{\prime}\div\tau)$. This correction is small although not insensible. This shows that the amount of change of obliquity has been slightly under-estimated. It does not, however, seem worth while to compute the corrected value for the change of obliquity in the integrations of the preceding paper.

The equation of conservation of moment of momentum, which is derived from the integration of the second of (71), clearly remains sensibly unaffected.

We see also from (70) that the time required for the changes has been over-estimated. If K_0, λ_0; K, λ be the initial and final values of K and λ at the beginning and end of one of the periods of integration; then it is obvious that our estimate of time should have been multiplied by some fraction lying between $1-\mathrm{K}_0(1-\lambda_0)^2$ and $1-\mathrm{K}(1-\lambda)^2$.

Now at the beginning of the first period $\mathrm{K}_0=\cdot0364$ and $\lambda_0=\cdot0365$, and at the end $\mathrm{K}=\cdot0865$ and $\lambda=\cdot0346$.

Whence $\mathrm{K}_0(1-\lambda_0)^2=\cdot034$, $\mathrm{K}(1-\lambda)^2=\cdot080$.

Hence it follows that the time, in the first period of the integration of Section 15, may have been over-estimated by some percentage less than some number lying between 3 and 8.

In fact, I have corrected the first period of that integration by a rather more tedious process than that here exhibited, and I found that the time was over-estimated by a little less than 3 per cent. And it was found that we ought to subtract from the 46,300,000 years comprised within the first period about 1,300,000 years. I also found that the error in the final value of the obliquity could hardly amount to more than 1′ or 2′.

In the later periods of integration the error in the time would no doubt be a little larger fraction of the time comprised within each period, but as it is not interesting to find the time in anything but round numbers, it is not worth while to find the corrections.

There is another point worth noticing. It might be suspected that when we approach the critical point where $n \cos i = 2\Omega$, where the rate of change of obliquity was found to vanish, the tidal movements might have become so rapid as seriously to affect the correctness of the tidal theory used; and accordingly it might be thought that the critical point was not reached even approximately when $n \cos i = 2\Omega$.

The preceding analysis will show at once that this is not the case. Near the critical point the solar terms have become negligeable; then if we put $\tau_{,}=0$ in the first of equations (69) we have

$$\frac{di}{dt}=\frac{1}{N}\cdot\frac{\tau^2}{\mathfrak{g}\omega_0}\sin 4\epsilon.\tfrac{1}{4}PQ[1-2\lambda\sec i+\mathrm{K}(1-2\lambda)(1-\tfrac{7}{4}\lambda+\lambda^2)].$$

The condition for the critical point in the first approximation was $2\lambda \sec i = 1$; if then i is so small that we may take $\sec i = 1$ in the inertia term, this condition also causes the inertia term to vanish.

Hence the corrected theory of tides makes no sensible difference in the critical point where $\frac{di}{dt}$ changes sign.

Having now disposed of these special points connected with previous results, I shall return to questions of general dynamics connected with the approximate solution of the forced vibrations of viscous spheroids; that is to say, I shall compare the results with those of—

*The forced oscillations of fluid spheroids.**

The same notation as before will serve again, and the equations of motion are

$$\left.\begin{array}{l} -\dfrac{dp}{dx}+\dfrac{d\mathrm{W}}{dx}-w\dfrac{d\alpha}{dt}=0 \\ \text{two similar equations} \\ \text{and } \dfrac{d\alpha}{dx}+\dfrac{d\beta}{dy}+\dfrac{d\gamma}{dz}=0 \end{array}\right\} \quad . \quad . \quad . \quad . \quad . \quad . \quad . \quad . \quad (73)$$

If the external tide-generating forces be those due to a potential per unit volume equal to $wr^i\mathrm{S}_i$, and $r=a+\sigma_i$ be the equation to the tidal spheroid, where S_i, σ_i are surface harmonics of the ith order, then we must put

$$\mathrm{W}=w\left(r^i\mathrm{S}_i+\frac{3g}{2i+1}\left(\frac{r}{a}\right)^i\sigma_i+(3a^2-r^2)\frac{g}{2a}\right),$$

the second term being the potential of the tidal protuberance, and the last of the mean sphere.

Differentiate the three equations of motion by x, y, z and add them, and we have

$$\nabla^2\left(p-w(3a^2-r^2)\frac{g}{2a}\right)=0.$$

* This is a slight modification of Sir W. THOMSON's investigation of the free oscillations of fluid spheres, Phil. Trans., 1863, p. 608.

Hence

$$p=w(3a^2-r^2)\frac{g}{2a}+\text{ solid harmonics } + \text{ a constant.}$$

Now when $r=a$, at the mean surface of the sphere, $p=gw\sigma_i$, therefore

$$p=w(a^2-r^2)\frac{g}{2a}+gw\sigma_i\left(\frac{r}{a}\right)^i.$$

Then substituting this value of p in the equations of motion (73),

$$\frac{du}{dt}=\frac{d}{dx}\left[r^i\mathrm{S}_i+\frac{3g}{2i+1}\sigma_i\left(\frac{r}{a}\right)^i+(3a^2-r^2)\frac{g}{2a}-(a^2-r^2)\frac{g}{2a}-g\sigma_i\left(\frac{r}{a}\right)^i\right]$$

$$\left.\begin{array}{l}\dfrac{d\alpha}{dt}=\dfrac{d}{dx}\left[r^i\mathrm{S}_i-\dfrac{2(i-1)}{2i+1}g\left(\dfrac{r}{a}\right)^i\sigma_i\right]\\ \text{and two similar equations}\end{array}\right\} \quad . \quad . \quad . \quad . \quad . \quad . \quad . \quad . \quad . \quad . \qquad (74)$$

The expression within brackets [ ] on the right is the effective disturbing potential, inclusive of the effects of mutual gravitation, and thus this process is exactly parallel to that adopted above in order to include the effects of mutual gravitation in the disturbing potential in the case of the viscous spheroid.

Now ρ, the radial velocity of flow, is equal to $\alpha\frac{x}{r}+\beta\frac{y}{r}+\gamma\frac{z}{r}$.

Therefore multiplying the equations (74) by $\frac{x}{r}, \frac{y}{r}, \frac{z}{r}$ and adding them, we have, by the properties of homogeneous functions,

$$\frac{d\rho}{dt}=i\left[r^{i-1}\mathrm{S}_i-\frac{2(i-1)}{2i+1}g\frac{r^{i-1}}{a^i}\sigma_i\right].$$

But when $r=a$, $\rho=\frac{d\sigma_i}{dt}$.

Therefore

$$\frac{d^2\sigma_i}{dt^2}=ia^{i-1}\mathrm{S}_i-\frac{2i(i-1)}{2i+1}\frac{g}{a}\sigma_i. \quad . \quad . \quad . \quad . \quad . \quad . \quad . \quad . \quad . \qquad (75)$$

Now suppose $\mathrm{S}_i=\mathrm{Q}_i\cos vt$, and that the tidal motion is steady, so that σ_i must be of the form $\mathrm{XQ}_i\cos vt$; then substituting in (75) this form of σ_i, we find

$$\mathrm{X}\left[-v^2+\frac{2i(i-1)}{2i+1}\frac{g}{a}\right]=ia^{i-1}.$$

Whence

$$\sigma_i=\frac{ia^{i-1}}{\frac{2i(i-1)}{2i+1}\frac{g}{a}-v^2}\mathrm{Q}_i\cos vt \qquad (76)$$

This gives the equation to the tidal spheroid.

Since the equilibrium tide, due to the disturbing potential, would be given by

$$\sigma_i=\frac{a^{i-1}}{\frac{2(i-1)}{2i+1}\frac{g}{a}}Q_i\cos vt,$$

it follows that inertia augments the height of tide in the proportion $1:1-\frac{(2i+1)}{2i(i-1)}\frac{a}{g}v^2$.

In the case where $i=2$, the augmentation is in the proportion $1:1-\frac{1}{2}\frac{v^2}{\mathfrak{g}}$.

We will now consider the nature of the motion by which each particle assumes its successive positions.

With the value of σ_i given in (76)

$$S_i-\frac{2(i-1)}{2i+1}\frac{g}{a^i}\sigma_i=\frac{-Q_iv^2}{\frac{2i(i-1)}{(2i+1)}\frac{g}{a}-v^2}\cos vt.$$

Then substituting in (74)

$$\left.\begin{aligned}&\frac{d\alpha}{dt}=-\frac{d}{dx}\frac{v^2\cos vt\,Q_ir^i}{\frac{2i(i-1)}{2i+1}\frac{g}{a}-v^2}\\&\text{and two similar equations}\end{aligned}\right\}\quad\ldots\ldots\ldots\quad(77)$$

Integrating with regard to t

$$\left.\begin{aligned}&\alpha=-\frac{d}{dx}\frac{Q_ir^iv\sin vt}{\frac{2i(i-1)}{2i+1}\frac{g}{a}-v^2}\\&\text{and two similar equations}\end{aligned}\right\}\quad\ldots\ldots\ldots\quad(78)$$

There might be a term introduced by integration, independent of the time, but this term must be zero, because if there were no disturbing force there would be no flow. Hence it is clear that there is a velocity potential ϑ, and that

$$\vartheta=\frac{1}{\frac{2i(i-1)}{2i+1}\frac{g}{a}-v^2}\frac{d}{dt}(r^iS_i)\quad\ldots\ldots\ldots\quad(79)$$

Now however slowly the motion takes place, there will always be a velocity potential, and if it be slow enough we may omit v^2 in the denominator of (79). In other words, if inertia be neglected the velocity potential is

$$\vartheta=\frac{2i+1}{2i(i-1)}\frac{a}{g}\frac{d}{dt}(r^iS_i).$$

For the sake of comparison with the approximate solution for the tides of a viscous spheroid, a precisely parallel process will now be carried out with regard to the fluid spheroid.

We obtain a first approximation for $\frac{d\alpha}{dt}$, when inertia is neglected, by omitting v^2 in the denominator of (77); whence

$$\frac{d\alpha}{dt}=-\frac{d}{dx}\left(\frac{2i+1}{2i(i-1)}\frac{g}{a}v^2\cos vt\,r^iQ_i\right).$$

Substituting this approximate value in the equations of motion (73) we have

$$\left.\begin{array}{c}-\frac{dp}{dx}+\frac{d}{dx}\left(W+w\frac{2i+1}{2i(i-1)}\frac{g}{a}v^2\cos vt\,r^iQ_i\right)=0\\ \text{and two similar equations}\end{array}\right\}\quad\ldots\ldots\quad(80)$$

From these equations it is obvious that the second approximation to the form of the tidal spheroid is found by augmenting the equilibrium tide due to the tide-generating potential $r^iQ_i\cos vt$ in the proportion $1+\frac{2i+1}{2i(i-1)}\frac{g}{a}v^2$ to unity.

When $i=2$ the augmenting factor is $1+\frac{1}{2}\frac{v^2}{\mathfrak{g}}$.

This is of course only an approximate result; the accurate value of the factor is $1\div\left(1-\frac{1}{2}\frac{v^2}{\mathfrak{g}}\right)$, and we see that the two agree if the squares and higher powers of $\frac{1}{2}\frac{v^2}{\mathfrak{g}}$ are negligeable.

Now in the case of the viscous tides we found the augmenting factor to be $1+\frac{79}{150}\frac{v^2}{\mathfrak{g}}\cos^2\epsilon$. When $\epsilon=0$, which corresponds to the case of fluidity, the expressions are closely alike, but we should expect that the 79 ought really to be 75.

The explanation which lies at the bottom of this curious discrepancy will be most easily obtained by considering the special case of a lunar semi-diurnal tide.

We found in Part II., equation (21), the following values for α, β, γ,

$$\left.\begin{array}{l}\alpha=\frac{w\tau}{38v}\sin 2\epsilon\left[(8a^2-5r^2)y+4x^2y\right]\\ \beta=\frac{w\tau}{38}\sin 2\epsilon\left[(8a^2-5r^2)x+4xy^2\right)\\ \gamma=\frac{w\tau}{38}\sin 2\epsilon\,.\,4xyz\end{array}\right\}\quad\ldots\ldots\quad(81)$$

where

$$\left.\begin{aligned} x&=r\sin\theta\cos(\phi-\omega t)\\ y&=r\sin\theta\sin(\phi-\omega t)\\ z&=r\cos\theta \end{aligned}\right\}.$$

Now consider the case when the viscosity is infinitely small: here ϵ is small, and $\sin 2\epsilon=\tan 2\epsilon=\frac{38v\omega}{5\mathfrak{g}wa^2}$.

Hence $\frac{\omega\tau}{38v}\sin 2\epsilon=\frac{\omega\tau}{5\mathfrak{g}a^2}$, which is independent of the viscosity.

By substituting this value in (81), we see that however small the viscosity, the nature of the motion, by which each particle assumes its successive positions, always preserves the same character; and the motion always involves molecular rotation.

But it has been already proved that, however slow the tidal motion of a fluid spheroid may be, yet the fluid motion is always irrotational.

Hence in the two methods of attacking the same problem, different first approximations have been used, whence follows the discrepancy of 79 instead of 75.

The fact is that in using the equations of flow of a viscous fluid, and neglecting inertia to obtain a first approximation, we postulate that $w\frac{d\alpha}{dt}$, $w\frac{d\beta}{dt}$, $w\frac{d\gamma}{dt}$, are less important than $v\nabla^2\alpha$, $v\nabla^2\beta$, $v\nabla^2\gamma$; and this is no longer the case if v be very small.

It does not follow therefore that, in approaching the problem of fluidity from the side of viscosity, we must necessarily obtain even an approximate result.

But the comparison which has just been made, shows that as regards the form of the tidal spheroid the two methods lead to closely similar results.

It follows therefore that, in questions regarding merely the form of the spheroid, and not the mode of internal motion, we only incur a very small error by using the limiting case when $v=0$ to give the solution for pure fluidity.

In the paper on "Precession" (Section 7), some doubt was expressed as to the applicability of the analysis, which gave the effects of tides on the precession of a rotating spheroid, to the limiting case of fluidity; but the present results seem to justify the conclusions there drawn.

The next point to be considered is the effects of inertia in—

The forced oscillations of an elastic sphere.

Sir WILLIAM THOMSON has found the form into which a homogeneous elastic sphere becomes distorted under the influence of a potential expressible as a solid harmonic of the points within the sphere. He afterwards supposed the sphere to possess the power of gravitation, and considered the effects by a synthetical method. The result is the equilibrium theory of the tides of an elastic sphere. When, however, the disturbing potential is periodic in time this theory is no longer accurate.

It has already been remarked that the approximate solution of the problem of determining the state of internal flow of a viscous spheroid when inertia is neglected, is identical in form with that which gives the state of internal strain of an elastic sphere; the velocities α, β, γ have merely to be read as displacements, and the coefficient of viscosity υ as that of rigidity.

The effects of mutual gravitation may also be introduced in both problems by the same artifice; for in both cases we may take, instead of the external disturbing potential $wr^2\mathrm{S}\cos vt$, an effective potential $wr^2\left(\mathrm{S}\cos vt-\mathfrak{g}\frac{\sigma}{a}\right)$, and then deem the sphere free of gravitational power.

Now Sir WILLIAM THOMSON's solution shows that the surface radial displacement (which is of course equal to σ) is equal to

$$\frac{5wa^3}{19\upsilon}\left(\mathrm{S}\cos vt-\mathfrak{g}\frac{\sigma}{a}\right) \quad . \quad . \quad . \quad . \quad . \quad . \quad . \quad . \quad (82)$$

If therefore we put (with Sir WILLIAM THOMSON) $\mathfrak{r}=\frac{19\upsilon}{5wa^2}$, we have $\frac{\sigma_{\prime}}{a}=\frac{\mathrm{S}}{\mathfrak{r}+\mathfrak{g}}\cos vt$.

This expression gives the equilibrium elastic tide, the suffix being added to the σ to indicate that it is only a first approximation.

Before going further we may remark that

$$\mathrm{S}\cos vt-\mathfrak{g}\frac{\sigma_{\prime}}{a}=\frac{\mathfrak{r}}{\mathfrak{r}+\mathfrak{g}}\mathrm{S}\cos vt \quad . \quad . \quad . \quad . \quad (83)$$

When we wish to proceed to a second approximation, including the effects of inertia, it must be noticed that the equations of motion in the two problems only differ in the fact that in that relating to viscosity the terms introduced by inertia are $-w\frac{d\alpha}{dt}$, $-w\frac{d\beta}{dt}$, $-\frac{d\gamma}{dt}$, whilst in the case of elasticity they are $-w\frac{d^2\alpha}{dt^2}$, $-w\frac{d^2\beta}{dt^2}$, $-w\frac{d^2\gamma}{dt^2}$. Hence a very slight alteration will make the whole of the above investigation applicable to the case of elasticity; we have, in fact, merely to differentiate the approximate values for α, β, γ twice with regard to the time instead of once.

Then just as before, we find the surface radial displacement, as far as it is due to inertia, to be (compare (55))

$$\frac{wv^2a^5}{\upsilon^2}\,\frac{79}{2.3.19^2}\,\frac{\mathrm{Y}}{r^2}\cos vt,$$

and $\frac{\mathrm{Y}}{wr^2}\cos vt$ must be put equal to (the first approximation) $\mathrm{S}\cos vt-\mathfrak{g}\frac{\sigma_{\prime}}{a}$. Hence by (57) and (83) the surface radial displacement due to inertia is $\frac{w^2v^2a^5}{\upsilon^2}\,\frac{79}{2.3.19^2}\,\frac{\mathfrak{r}}{\mathfrak{r}+}\mathrm{S}\cos vt$.

To this we must add the displacement due directly to the effective disturbing potential $wr^2\left(\mathrm{S}\cos vt-\mathfrak{g}\frac{\sigma}{a}\right)$, where σ is now the second approximation. This we know from (82) is equal to

$$\frac{5wa^3}{19\nu}\left(\mathrm{S}\cos vt-\mathfrak{g}\frac{\sigma}{a}\right).$$

Hence the total radial displacement is

$$\frac{5wa^3}{19\nu}\left(\mathrm{S}\cos vt-\mathfrak{g}\frac{\sigma}{a}+\frac{5wa^2}{19\nu}\cdot\frac{79v^2}{150}\,\frac{\mathfrak{r}}{\mathfrak{r}+\mathfrak{g}}\mathrm{S}\cos vt\right).$$

But the total radial displacement is itself equal to σ.

Therefore

$$\mathfrak{r}\frac{\sigma}{a}=\mathrm{S}\cos vt-\mathfrak{g}\frac{\sigma}{a}+\frac{79v^2}{150(\mathfrak{r}+\mathfrak{g})}\mathrm{S}\cos vt,$$

and

$$\frac{\sigma}{a}=\frac{\mathrm{S}}{\mathfrak{r}+\mathfrak{g}}\cos vt\left(1+\frac{79v^2}{150(\mathfrak{r}+\mathfrak{g})}\right).$$

This is the second approximation to the form of the tidal spheroid, and from it we see that inertia has the effect of increasing the ellipticity of the spheroid in the proportion $1+\frac{79v^2}{150(\mathfrak{r}+\mathfrak{g})}$.

Analogy with (76) would lead one to believe that the period of the gravest vibration of an elastic sphere is $2\pi\left(\frac{79}{150\mathfrak{r}}\right)^{\frac{1}{2}}$; this result might be tested experimentally.

If $\mathfrak{g}$ be put equal to zero, the sphere is devoid of gravitation, and if $\mathfrak{r}$ be put equal to zero the sphere becomes perfectly fluid; but the solution is then open to objections similar to those considered, when viscosity graduates into fluidity.

It is obvious that the whole of this present part might be easily adapted to that hypothesis of elastico-viscosity which was considered in the paper on "Tides," but it does not at present seem worth while to do so.

By substituting these second approximations in the equations of motion again, we might proceed to a third approximation, and so on; but the analytical labour of the process would become very great.

IV. *Discussion of the applicability of the results to the history of the earth.*

The first paper of this series was devoted to the consideration of inequalities of short period, in the state of flow of the interior, and in the form of surface, produced in a rotating viscous sphere by the attraction of an external disturbing body: this was the theory of tides. The investigation was admitted to be approximate from two causes —(i) the neglect of the inertia of the relative motion of the parts of the spheroid; (ii) the neglect of tangential action between the surface of the mean sphere and the tidal protuberances.

In the second paper the inertia was still neglected, but the effects of these tangential actions were considered, in as far as they modified the rotation of the spheroid as a whole. In that paper the sphere was treated as though it were rigid, but had rigidly attached to its surface certain inequalities, which varied in distribution from instant to instant according to the tidal theory.

In order to justify this assumption, it is now necessary to examine whether the tidal protuberances may be regarded as instantaneously and rigidly connected with the rotating sphere. If there is a secular distortion of the spheroid in excess of the regular tidal flux and reflux, the assumption is not rigorously exact; but if the distortion be very slow, the departure from exactness may be regarded as insensible.

The first problem in the present paper is the investigation of the amount of secular distortion, and it is treated only in the simple case of a single disturbing body, or moon, moving in the equator of the tidally-distorted spheroid or earth.

It is found, then, that the form of the lagging tide in the earth is not such that the pull, exercised by the moon on it, can retard the earth's rotation exactly as though the earth were a rigid body. In other words, there is an unequal distribution of the tidal frictional couple in various latitudes.

We may see in a general way that the tidal protuberance is principally equatorial, and that accordingly the moon tends to retard the diurnal rotation of the equatorial portions of the sphere more rapidly than that of the polar regions. Hence the polar regions tend to outstrip the equator, and there is a slow motion from west to east relatively to the equator.

When, however, we come to examine numerically the amount of this screwing motion of the earth's mass, it appears that the distortion is exceedingly slow, and accordingly the assumption of the instantaneous rigid connexion of the tidal protuberance with the mean sphere is sufficiently accurate to allow all the results of the paper on "Precession" to hold good.

In the special case, which was the subject of numerical solution in that paper, we were dealing with a viscous mass which in ordinary parlance would be called a solid, and it was maintained that the results might possibly be applicable to the earth within the limits of geological history.

Now the present investigation shows that if we look back 45,000,000 years from the present state of things, we might find a point in lat. 30° further west with reference to a point on the equator, by $4\frac{3}{4}'$ than at present, and a point in lat. 60° further west by $14\frac{1}{4}'$. The amount of distortion of the surface strata is also shown to be exceedingly minute.

From these results we may conclude that this cause has had little or nothing to do with the observed crumpling of strata, at least within recent geological times.

If, however, the views maintained in the paper on "Precession" as to the remote history of the earth are correct, it would not follow, from what has been stated above, that this cause has never played an important part; for the rate of the screwing of the earth's mass varies inversely as the sixth power of the moon's distance, multiplied by the angular velocity of the earth relatively to the moon. And according to that theory, in very early times the moon was very near the earth, whilst the relative angular velocity was comparatively great. Hence the screwing action may have been once sensible.*

Now this sort of motion, acting on a mass which is not perfectly homogeneous, would raise wrinkles on the surface which would run in directions perpendicular to the axis of greatest pressure.

In the case of the earth the wrinkles would run north and south at the equator, and would bear away to the eastward in northerly and southerly latitudes; so that at the north pole the trend would be north-east, and at the south pole north-west. Also the intensity of the wrinkling force varies as the square of the cosine of the latitude, and is thus greatest at the equator, and zero at the poles. Any wrinkle when once formed would have a tendency to turn slightly, so as to become more nearly east and west, than it was when first made.

The general configuration of the continents (the large wrinkles) on the earth's surface appears to me remarkable when viewed in connexion with these results.

There can be little doubt that, on the whole, the highest mountains are equatorial, and that the general trend of the great continents is north and south in those regions. The theoretical directions of coast line are not so well marked in parts removed from the equator.

* This result is not strictly applicable to the case of infinitely small viscosity, because it gives a finite though very small circulation, if the coefficient of viscosity be put equal to zero.

By putting $\epsilon=0$ in (17′), Part I., we find a superior limit to the rate of distortion. With the present angular velocities of the earth and moon, $\frac{d\text{L}}{dt}$ must be less than $5\times10^{-9}\cos^2\theta$ in degrees per annum.

It is easy to find when $\frac{d\text{L}}{dt}$ would be a maximum in the course of development considered in "Precession;" for, neglecting the solar effects, it will be greatest when $\tau^2(n-\Omega)$ is greatest.

Now $\tau^2(n-\Omega)$ varies as $[1+\mu-\mu\xi-\frac{\Omega_0}{n_0}.\xi^{-3}]\xi^{-12}$, and this function is a maximum when

$$\xi^{-4}-\frac{12}{15}(1+\mu)\frac{n_0}{\Omega_0}\xi^{-1}+\frac{11}{15}\mu\frac{n_0}{\Omega_0}=0.$$

Taking $\mu=4\cdot0074$, and $\frac{n_0}{\Omega_0}=27\cdot32$, we have $\xi^{-4}-109\cdot45\xi^{-1}+80\cdot293=0$.

The solution of this is $\xi=\cdot2218$.

With this solution $\frac{d\text{L}}{dt}$ will be found to be 56 million times as great as at present, being equal to $18'\cos^2\theta$ per annum. With this value of ξ, the length of the day is 5 hours 50 minutes, and of the month 7 hours 10 minutes.

This gives a superior limit to the greatest rate of distortion which can ever have occurred.

By (19′), however, we see that the rate of distortion per unit increment of the moon's distance may be made as large as we please by taking the coefficient of viscosity small enough.

These considerations seem to show that there is no reason why this screwing action of the earth should not once have had considerable effects. (Added October 15, 1879.)

The great line of coast running from North Africa by Spain to Norway has a decidedly north-easterly bearing, and the long Chinese coast exhibits a similar tendency. The same may be observed in the line from Greenland down to the Gulf of Mexico, but here we meet with a very unfavourable case in Panama, Mexico, and the long Californian coast line.

From the paucity of land in the southern hemisphere the indications are not so good, nor are they very favourable to these views. The great line of elevation which runs from Borneo through Queensland to New Zealand might perhaps be taken as an example of north-westerly trend. The Cordilleras run very nearly north and south, but exhibit a clear north-westerly twist in Tierra del Fuego, and there is another slight bend of the same character in Bolivia.

But if this cause was that which principally determined the direction of terrestrial inequalities, then the view must be held that the general position of the continents has always been somewhat as at present, and that, after the wrinkles were formed, the surface attained a considerable rigidity, so that the inequalities could not entirely subside during the continuous adjustment to the form of equilibrium of the earth, adapted at each period to the lengthening day. With respect to this point, it is worthy of remark that many geologists are of opinion that the great continents have always been more or less in their present positions.

An inspection of Professor SCHIAPPARELLI's map of Mars,* I think, will prove that the north and south trend of continents is not something peculiar to the earth. In the equatorial regions we there observe a great many very large islands, separated by about twenty narrow channels running approximately north and south. The northern hemisphere is not given beyond lat. 40°, but the coast lines of the southern hemisphere exhibit a strongly marked north-westerly tendency. It must be confessed, however, that the case of Mars is almost too favourable, because we have to suppose, according to the theory, that its distortion is due to the sun, from which the planet must always have been distant. The very short period of the inner satellite shows, however, that the Martian rotation must have been (according to the theory) largely retarded; and where there has been retardation, there must have been internal distortion.

The second problem which is considered in the first part of the present paper is concerned with certain secondary tides. My attention was called to these tides by some remarks of Dr. JULES CARRET,† who says :—

"Les actions perturbatrices du soleil et de la lune, qui produisent les mouvements coniques de la précession des équinoxes et de la nutation, n'agissent que sur cette portion de l'ellipsoïde terrestre qui excède la sphère tangente aux deux pôles, c'est-à-dire, en admettant l'état pâteux de l'intérieur, à peu près uniquement sur ce

* 'Appendice alle Memorie della Società degli Spettroscopisti Italiani,' 1878, vol. vii., for a copy of which I have to thank M. SCHIAPPARELLI.

† Société Savoisienne d'Histoire et d'Archéologie, May 23, 1878. He is also author of a work, 'Le Déplacement Polaire.' I think Dr. CARRET has misunderstood Mr. EVANS.

que l'on est convenu d'appeler la croûte terrestre, et presque sur toute la croûte terrestre. La croûte glisse sur l'intérieur plastique. Elle parvient à entrainer l'intérieur, car, sinon, l'axe de la rotation du globe demeurerait parallèle à lui-même dans l'espace, ou n'éprouverait que des variations insignifiantes, et le phenomène de la précession des equinoxes n'existerait pas. Ainsi la croûte et l'intérieur se meuvent de quantités inégales, d'où le déplacement géographique du pôle sur la sphère.

"Cette idée a été émise, je crois, pour la première fois, par M. EVANS; depuis par M. J. PÉROCHE."

Now with respect to this view, it appears to me to be sufficient to remark that, as the axes of the precessional and nutational couples are fixed relatively to the moon, whilst the earth rotates, therefore the tendency of any particular part of the crust to slide over the interior is reversed in direction every twelve lunar hours, and therefore the result is not a secular displacement of the crust, but a small tidal distortion.

As, however, it was just possible that this general method of regarding the subject overlooked some residual tendency to secular distortion, I have given the subject a more careful consideration. From this it appears that there is no other tendency to distortion besides that arising out of tidal friction, which has just been discussed. It is also found that the secondary tides must be very small compared with the primary ones; with the present angular velocity of diurnal rotation, probably not so much in height as one-hundredth of the primary lunar semi-diurnal bodily tide.

It seems out of the question that any heterogeneity of viscosity could alter this result, and therefore it may, I think, be safely asserted that any sliding of the crust over the interior is impossible—at least as arising from this set of causes.

The second part of the paper is an investigation of the amount of work done in the interior of the viscous sphere by the bodily tidal distortion.

According to the principles of energy, the work done on any element makes itself manifest in the form of heat. The whole work which is done on the system in a given time is equal to the whole energy lost to the system in the same time. From this consideration an estimate was given, in the paper on "Precession," of the whole amount of heat generated in the earth in a given time. In the present paper the case is taken of a moon moving round the earth in the plane of the equator, and the work done on each element of the interior is found. The work done on the whole earth is found by summing up the work on each element, and it appears that the work per unit time is equal to the tidal frictional couple multiplied by the relative angular velocity of the two bodies. This remarkably simple law results from a complex law of internal distribution of work, and its identity with the law found in "Precession," from simple considerations of energy, affords a valuable confirmation of the complete consistency of the theory of tides with itself.

Fig. 2 gives a graphical illustration of the distribution in the interior of the work done, or of the heat generated, which amounts to the same thing. The reader is referred to Part II. for an explanation of the figure. Mere inspection of the figure shows that by far the larger part of the heat is generated in the central parts, and calculation shows that about one-third of the whole heat is generated within the central one-eighth of the volume, whilst in a spheroid of the size of the earth only one-tenth is generated within 500 miles of the surface.

In the paper on "Precession" the changes in the system of the sun, moon, and earth were traced backwards from the present lengths of day and month back to a common length of day and month of 5 hours 36 minutes, and it was found that in such a change heat enough must have been generated within the earth to raise its whole mass 3000° Fahr. if applied all at once, supposing the earth to have the specific heat of iron. It appeared to me at that time that, unless these changes took place at a time very long antecedent to geological history, then this enormous amount of internal heat generated would serve in part to explain the increase of temperature in mines and borings. Sir WILLIAM THOMSON, however, pointed out to me that the distribution of heat-generation would probably be such as to prevent the realisation of my expectations. I accordingly made the further calculations, connected with the secular cooling of the earth, comprised in the latter portion of Part II.

It is first shown that, taking certain average values for the increase of underground temperature and for the conductivity of the earth, then the earth (considered homogeneous) must be losing by conduction outwards an amount of energy equal to its present kinetic energy of rotation in about 262 million years.

It is next shown that in the passage of the system from a day of 5 hours 40 minutes to one of 24 hours, there is lost to the system an amount of energy equal to $13\frac{1}{2}$ times the present kinetic energy of rotation of the earth. Thus it appears that, at the present rate of loss, the internal friction gives a supply of heat for 3,560 million years. So far it would seem that internal friction might be a powerful factor in the secular cooling of the earth, and the next investigation is directly concerned with that question.

In the case of the tidally-distorted sphere the distribution of heat-generation depends on latitude as well as depth from the surface, but the average law of heat-generation, as dependent on depth alone, may easily be found. Suppose, then, that we imagine an infinite slab of rock 8,000 miles thick, and that we liken the medial plane to the earth's centre and suppose the heat to be generated uniformly in time, according to the average law above referred to. Then conceive the two faces of the slab to be always kept at the same constant temperature, and that initially, when the heat-generation begins, the whole slab is at this same temperature. The problem then is, to find the rate of increase of temperature going inwards from either face of the slab after any time.

This problem is solved, and by certain considerations (for which the reader is referred back) is made to give results which must agree pretty closely with the temperature gradient at the surface of an earth in which $13\frac{1}{2}$ times the present kinetic energy of earth's rotation, estimated as heat, is uniformly generated in time, with the average space distribution referred to. It appears that at the end of the heat-generation the temperature gradient at the surface is sensibly the same, at whatever rate the heat is generated, provided it is all generated within 1,000 million years; but the temperature gradient can never be quite so steep as if the whole heat were generated instantaneously. The gradient, if the changes take place within 1,000 million years, is found to be about 1° Fahr. in 2,600 feet. Now the actually observed increase of underground temperature is something like 1° Fahr. in 50 feet; it therefore appears that perhaps one-fiftieth of the present increase of underground temperature may possibly be referred to the effects of long past internal friction. It follows, therefore, that Sir WILLIAM THOMSON'S investigation of the secular cooling of the earth is not sensibly affected by these considerations.

If at any time in the future we should attain to an accurate knowledge of the increase of underground temperature, it is just within the bounds of possibility that a smaller rate of increase of temperature may be observed in the equatorial regions than elsewhere, because the curve of equal heat generation, which at the equator is nearly 500 miles below the surface, actually reaches the surface at the pole.

The last problem here treated is concerned with the effects of inertia on the tides of a viscous spheroid. As this part will be only valuable to those who are interested in the actual theory of tides, it may here be dismissed in a few words. The theory used in the two former papers, and in the first two parts of the present one, was founded on the neglect of inertia; and although it was shown in the paper on "Tides" that the error in the results could not be important, in the case of a sphere disturbed by tides of a frequency equal to the present lunar and solar tides, yet this neglect left a defect in the theory which it was desirable to supply. Moreover it was possible that, when the frequency of the tides was much more rapid than at present (as was found to have been the case in the paper on "Precession"), the theory used might be seriously at fault.

It is here shown (see (62)) that for a given lag of tide the height of tide is a little greater, and that for a given frequency of tide the lag is a little greater than the approximate theory supposed.

A rough correction is then applied to the numerical results given in the paper on "Precession" for the secular changes in the configuration of the system; it appears that the time occupied by the changes in the first solution (Section 15) is overstated by about one-fortieth part, but that all the other results, both in this solution and the other, are left practically unaffected. To the general reader, therefore, the value of this part of the paper simply lies in its confirmation of previous work.

From a mathematical point of view, a comparison of the methods employed with those for finding the forced oscillations of fluid spheres is instructive.

Lastly, the analytical investigation of the effects of inertia on the forced oscillations of a viscous sphere is found to be applicable, almost verbatim, to the same problem concerning an elastic sphere. The results are complementary to those of Sir WILLIAM THOMSON'S statical theory of the tides of an elastic sphere.

www.ingramcontent.com/pod-product-compliance
Lightning Source LLC
Chambersburg PA
CBHW022018190326
41519CB00010B/1554
9783337213992